AF877084

Journal of
Mathematical
Methods
in Biosciences

BIOMETRICAL JOURNAL

Volume 27 · 1985 · Number 2 March

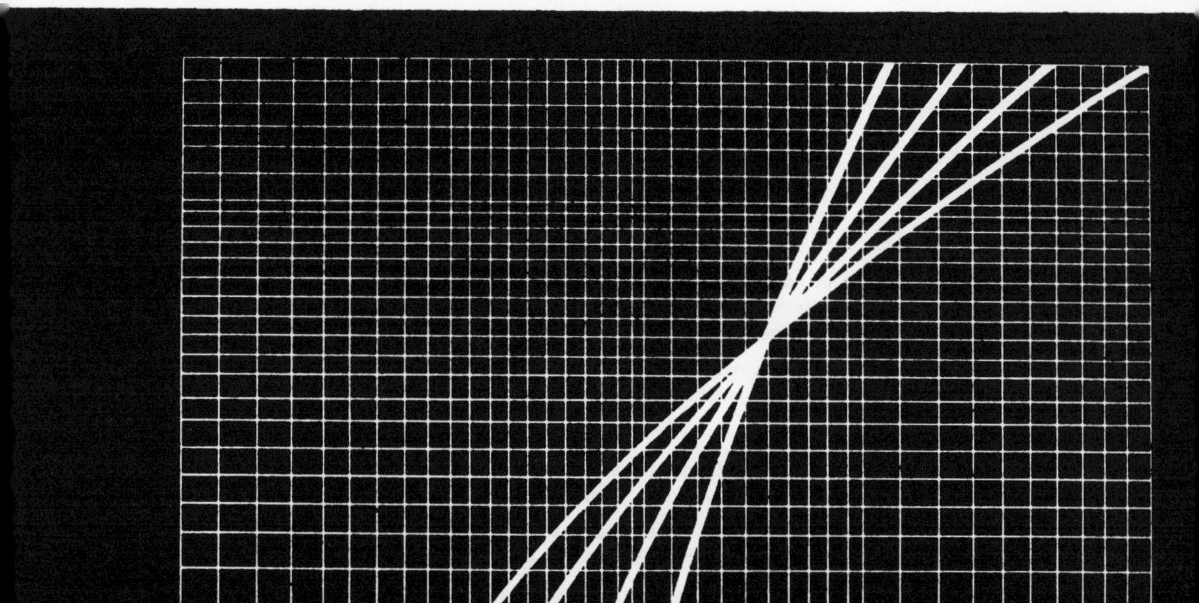

Akademie-Verlag · Berlin
and
VCH Publishers, Inc. · Deerfield Beach/Florida

ISSN 0323-3847 Biom. J., Berlin 27 (1985) 2, 121–240

Information on the Journal

Biometrical Journal was founded for the purpose of publishing contributions on mathematical penetration of biosciences.

The scope includes papers on new theoretical aspects of mathematics and its application to biological sciences in the widest sense (including biolcgy, medicine, agricultural science, forestry) or on the application of known mathematical and statistical methods to new areas. These may be methods of mathematical statistics and approaches to mathematico-cybernetik model construction of biological systems with due consideration of electronic data processing.

Accepted for publication: original papers, summary reports on the latest developments in the above areas, proceedings and book reviews.

Mailing address:
Institut für Mathematik der AdW
DDR - 1086 Berlin
Mohrenstraße 39

Instructions for authors:

1. Manuscripts should be submitted in two copies typed with double line spacing on one side of the page and with sufficient margin. Authors are expected to keep one complete copy of their paper for themselves. A manuscript should not be longer than 30 typewritten pages.
2. The institute at which the work on the paper was done should be listed below the name of the author (with the initials of his Christian names) below the title.
3. The full address of the author must be listed at the end of the paper.
4. Each original paper should include an abstract and two summaries, one in the language of the paper and the other in one of the publication languages (German, English, French, Russian).
5. Illustrations and captions must be enclosed separately and they should be continuously numbered. The number of illustrations should be kept as small as possible.
6. Special markings in colour (e.g. for formulae) should be avoided.
7. Smaller type for a section of the text should be denoted by a vertical stroke drawn at the left on the section concerned and by writing the instruction "smaller type" in the margin space next to the stroke.
8. Literature in the text should be quoted together with the name of the author and the year of publication. Each paper has to be completed by a list of references specifying the name and initials of the Christian names of the author, year, title of paper, name of journal, volume, issue, and for books also the place of publication.
9. Corrections or changes of the paper must be avoided after the manuscript has been typeset. The publishers reserve the right to charge the author for corrections that exceed ten per cent of the overall cost of typesetting.
10. Each author is entitled to receive 50 reprints free of charge and up to 100 additional copies on payment.

BIOMETRICAL JOURNAL

BIOMETRISCHE ZEITSCHRIFT

Journal of
Mathematical Methods
in Biosciences

Zeitschrift für
mathematische Methoden
in den Biowissenschaften

Volume 27
1985
Number 2

Edited at Institute
of Mathematics
Academy of Sciences of the G.D.R.
by
Erna Weber (Berlin), Editor-in-Chief

Editors
Heinz Ahrens (Berlin), Klaus Bellmann (Berlin)

Editorial Board
J. Adam (Halle), G. Enderlein (Berlin),
D. J. Finney (Edinburgh), D. Firescu (Bucharest),
B. W. Gnedenko (Moscow), S. Koller (Mainz),
H. L. Le Roy (Zurich), C. C. Li (Pittsburgh),
G. A. Lienert (Nuremberg), A. Linder (Geneva),
L. Martin (Brussels), W. Oehmisch (Berlin),
T. Postelnicu (Bucharest), D. Rasch (Rostock),
V. A. Ratner (Novosibirsk), J.-H. Scharf (Halle),
L. Schmetterer (Vienna), D. Stoyan (Freiberg),
P. Tautu (Heidelberg), R. Trommer (Eberswalde)

AKADEMIE-VERLAG · BERLIN
AND
VCH PUBLISHERS, INC
DEERFIELD BEACH/FLORIDA

Inquiries, subscriptions, and orders for single copies from customers in the USA and Canada should be directed to

VCH Publishers, Inc.,
303 N. W., 12th Avenue, Deerfield Beach, FL 33442–1705, USA;

from all other countries to
Akademie-Verlag, Leipziger Str. 3–4, DDR - 1086 Berlin.

Application to mail at second class postage paid at Jamaica, NY, 11431.
Airfreighted and mailed in the United States by Publications Expediting, Inc., 200 Meacham Avenue, Elmont, NY 11003.

USA Postmasters: Send address changes to Biometrical Journal to VCH Publishers, Inc., 303 N.W. 12th Avenue, Deerfield Beach, FL 33442–1705, USA.

Published 1 volume with 8 issues per year.

(February, March, May, June, August, September, November, December)
by Akademie-Verlag Berlin, GDR.

Annual subscription price: U.S. $160. – including postage.

Editorial notes:

Manuscripts should be addressed to the editorial board:
Institut für Mathematik der AdW, DDR - 1086 Berlin, Mohrenstraße 39.
Manuscripts should be submitted preferentially in English.
Besides papers in German, French and Russian are admitted.
Authors are held fully responsible for all information contained in their papers.

Biometrial Journal

Herausgeber: Prof. Dr. habil. Dr. h. c. Erna Weber, Institut für Mathematik der AdW, DDR - 1086 Berlin, Mohrenstraße 39.

Verlag: Akademie-Verlag, DDR - 1086 Berlin, Leipziger Straße 3–4; Fernruf: 2 23 62 01 oder 2 23 62 29; Telex-Nr.: 11 44 20; Bank: Staatsbank der DDR, Berlin, Kto-Nr.: 6826-26-20712.

Chefredaktion: Prof. Dr. habil. H. Ahrens und Dr. habil. K. Bellmann.

Anschrift der Redaktion: Institut für Mathematik der AdW, DDR - 1086 Berlin, Mohrenstraße 39.

Veröffentlicht unter der Lizenznummer 1335 des Presseamtes beim Vorsitzenden des Ministerrates der Deutschen Demokratischen Republik.

Gesamtherstellung: VEB Druckerei »Gottfried Wilhelm Leibniz«, DDR - 4450 Gräfenhainichen.

Erscheinungsweise: Die Zeitschrift erscheint jährlich in einem Band mit 8 Heften.

Bestellnummer dieses Heftes: 1060/27/2.

Urheberrecht: Alle Rechte vorbehalten insbesondere die der Übersetzung. Kein Teil dieser Zeitschrift darf in irgendeiner Form – durch Photokopie, Mikrofilm oder irgendein anderes Verfahren – ohne schriftliche Genehmigung des Verlages reproduziert oder in eine von Maschinen, insbesondere von Datenverarbeitungsanlagen verwendbare Sprache übertragen oder übersetzt werden.

All rights reserved (including those of translation into foreign languages). No part of this issue may be reproduced in any form by photoprint, microfilm or any other means, nor transmitted or translated into a machine language, without written permission from the publishers.

© 1985 by Akademie-Verlag Berlin. Printed in the German Democratic Republic.

Biom. J. 27 (1985) 2, 123–138

Robust Univariate Two-Way Classification[1]

N. BALAKRISHNAN and M. L. TIKU

Department of Mathematical Sciences
McMaster University, Hamilton, Ontario, Canada

A. H. El. SHAARAWI

National Water Research Institute, Ontario

Abstract

We develop a robust classification procedure, based on TIKU's (1967, 1980, 1982) MML (modified maximum likelihood) estimators, for classifying an observation in one of the two populations π_1 and π_2 and show that this procedure is superior to the classical and nonparametric procedures.

1. Introduction

Let x_{1i}, $i=1, 2, \ldots, n_1$, and x_{2i}, $i=1, 2, \ldots, n_2$, be independent random samples from populations π_1 and π_2, respectively; we assume that π_1 and π_2 are identical (other than a location shift) and have finite means and a common finite variance. On the basis of these two samples, one wants to classify a new independent observation x_0 in one of the two populations π_1 and π_2. Let e_1 be the probability of classifying x_0 in π_2 when in fact it has come from π_1; and e_2 be the probability of classifying x_0 in π_1 when in fact it has from π_2; e_1 and e_2 are called errors of misclassification. We develop a classification procedure based on TIKU's (1967, 1980, 1982) MML estimators, analogous to the classical procedure based on sample means and variances, and show that this new procedure is superior to the classical procedure, superior in the sense that for a given e_1 this new procedure has higher values of $1-e_2$ (the probability of correctly classifying x_0 in π_2), for numerous non-normal populations. We also show that this new procedure is superior to a number of nonparametric procedures in having smaller error rates $(e_1+e_2)/2$.

2. Classical Procedure

Let x_{1i}, $i=1, 2, \ldots, n_1$, and x_{2i}, $i=1, 2, \ldots, n_2$, be independent random samples from populations π_1 and π_2 assumed to be normal $N(\mu_1, \sigma)$ and $N(\mu_2, \sigma)$, respectively. Write

(2.1) $$\bar{x}_1=\sum_{i=1}^{n_1} x_{1i}/n \quad \text{and} \quad \bar{x}_2=\sum_{i=1}^{n_2} x_{2i}/n_2$$

[1] This is an invited paper.

and

$$s^2=\{(n_1-1)\,s_1^2+(n_2-1)\,s_2^2\}/(n_1+n_2-2)\ ;$$

$s_1^2=\sum_{i=1}^{n_1}(x_{1i}-\bar{x}_1)^2/(n_1-1)$ and $s_2^2=\sum_{i=1}^{n_2}(x_{2i}-\bar{x}_2)^2/(n_2-1)$. To classify a new independent observation x_0, the classical procedure is to classify x_0 in π_1 or π_2 according as (ANDERSON, 1951)

$$V \gtrless C\,, \tag{2.2}$$

where the constant C is so chosen that one of the errors of misclassification (say, e_1) assumes a pre-specified level, usually 0.05 or 0.01;

$$V=\{x_0-(\bar{x}_1+\bar{x}_2)/2\}\,(\bar{x}_1-\bar{x}_2)/s^2\,. \tag{2.3}$$

Although $\bar{x}_1$ and $\bar{x}_2$ are independent normals $N(\mu_1,\ \sigma/\sqrt{n_1})$ and $N(\mu_2,\ \sigma/\sqrt{n_2})$, respectively, and $(n_1+n_2-2)\,s^2/\sigma^2$ is an independent chi-square, but the distribution of V does not assume a manageable form (see SINHA and GIRI, 1975, for example). It is, therefore, difficult to evaluate the constant C for a given e_1 (or e_1 for a given C) and $1-e_2$ (the probability of correctly classifying x_0 in π_2), even numerically. However, the first four moments of V can be worked out and these are given in Appendix *A*. Using JOHNSON et al. (1963) tables, four-moment Pearson curves can be fitted to these four moments and approximate values of C obtained.

In the asymptotic situation (n_1 and n_2 both tend to infinity), however, $\bar{x}_1$ and $\bar{x}_2$ tend to their expected values $E(\bar{x}_1)=\mu_1$ and $E(\bar{x}_2)=\mu_2$ and s tends to its expected value $E(s)=\sigma$ in which case the distribution of V is the same as that of Fisher linear discriminant function

$$U=\{x_0-(\mu_1+\mu_2)/2\}\,(\mu_1-\mu_2)/\sigma^2\,. \tag{2.4}$$

The distribution of U is clearly normal with mean

$$E(U \mid x_0\in\pi_1)=\delta/2 \quad\text{and}\quad E(U \mid x_0\in\pi_2)=-\delta/2$$

and variance

$$\mathrm{Var}\,(U)=\delta\ ; \tag{2.5}$$

$\delta=(\mu_1-\mu_2)^2/\sigma^2$ is the standardized distance between the two populations π_1 and π_2.

Under the assumption of normality, the asymptotic value of C is given by

$$C=(\delta/2)-1.64\,\sqrt{\delta}, \qquad (e_1=0.05)\,. \tag{2.6}$$

However, C cannot be determined since δ is not known. In practice, therefore, C is replaced by a suitable estimate

$$\hat{C}=(\hat{\delta}/2)-1.64\,\sqrt{\hat{\delta}}\,. \tag{2.7}$$

Realizing that t^2,

$$t=(\bar{x}_1-\bar{x}_2)/s\ \sqrt{\left(\frac{1}{n_1}+\frac{1}{n_2}\right)}\,, \tag{2.8}$$

has a noncentral F-distribution with $(1, n_1+n_2-2)$ degrees of freedom and noncentrality parameter δ, an unbiassed estimator of δ is given by

$$\tilde{\delta}-\left(\frac{1}{n_1}+\frac{1}{n_2}\right), \tag{2.9}$$

where (LACHENBRUCH and MICKEY, 1968)

$$\tilde{\delta}=\{(n_1+n_2-4)/(n_1+n_2-2)\}\,(\bar{x}_1-\bar{x}_2)^2/s^2\,. \tag{2.10}$$

Since (2.9) can assume negative values which are not permissible, $\tilde{\delta}$ itself provides a suitable estimator of δ.

Thus, the classical procedure is to classify x_0 in π_1 or π_2 according as

$$V \gtrless \tilde{C}\,, \tag{2.11}$$

where $\tilde{C}$ is given by (2.7) and (2.10); this classification procedure has $e_1 \simeq 0.05$ (approximately), for large n_1 and n_2.

In practice, however, the underlying distribution will hardly ever be exactly normal (in fact, the underlying distribution will hardly ever be known exactly) and that raises the question whether the above classification procedure is robust to departures from normality. We will see later that it is not, for most non-normal distributions prevalent in practice.

3. Robust Classification Procedure

A robust classification procedure may be obtained by replacing the sample mean and standard deviation $\bar{x}$ and s by TIKU's (1967, 1980, 1982) MML estimators $\hat{\mu}$ and $\hat{\sigma}$; the expressions for $\hat{\mu}$ and $\hat{\sigma}$ are given in Appendix B. Thus a robust classification procedure may be to classify x_0 in π_1 or π_2 according as

$$V_R \gtrless C^*\,, \tag{3.1}$$

where

$$V_R=\{x_0-(\hat{\mu}_1+\hat{\mu}_2)/2\}\,(\hat{\mu}_1-\hat{\mu}_2)/\hat{\sigma}^2\,; \tag{3.2}$$

$\hat{\sigma}^2=\{(A_1-1)\,\hat{\sigma}_1^2+(A_2-1)\,\hat{\sigma}_2^2\}/(A_1+A_2-2)$, and C^* is so chosen that e_1 is equal to a pre-specified level (0.05, say); $\hat{\mu}_1$ and $\hat{\sigma}_1$ are the MML estimators calculated from the first sample x_{1i}, $i=1, 2, \ldots, n_1$, and $\hat{\mu}_2$ and $\hat{\sigma}_2$ are the MML estimators calculated from the second sample x_{2i}, $i=1, 2, \ldots, n_2$.

Theorem 1: Under the assumption of normality, the distribution of V_R is exactly the same as that of V, for large $A_1=n_1-2r_1$ and $A_2=n_2-2r_2$.

Proof: The result follows immediately from the fact that for large A, the MML estimators $\hat{\mu}$ and $\hat{\sigma}$ have exactly the same distributions as the sample mean and standard deviation $\bar{x}$ and s, namely, $\sqrt{m}\,(\hat{\mu}-\mu)/\sigma$, $(m=n-2r+2r\beta$; Appendix B) is a normal $N(0, 1)$ and $(A-1)\,\hat{\sigma}^2/\sigma^2$ is an independent chi-square with $A-1$ degrees of freedom; for the proof, see TIKU (1982a, pp. 626, 627). For large A_1 and A_2, therefore, the estimators $\hat{\mu}_1$, $\hat{\mu}_2$ and $\hat{\sigma}$ have exactly the same distributional properties as $\bar{x}_1$, $\bar{x}_2$ and s, respectively, and the result follows.

In practice C^* is replaced by a suitable estimate, and a robust (hopefully) procedure is to classify x_0 in π_1 or π_2 according as

(3.3) $$V_R \gtrless \hat{C}^* ,$$

where

(3.4) $$\hat{C}^* = (\hat{\delta}^*/2) - 1.64 \sqrt{\hat{\delta}^*}, \qquad (e_1 \simeq 0.05) ;$$
$$\hat{\delta}^* = \{(A_1 + A_2 - 4)/(A_1 + A_2 - 2)\} (\hat{\mu}_1 - \hat{\mu}_2)^2/\hat{\sigma}^2 .$$

Theorem 2: Under the assumption of normality, the V- and V_R-procedures have the same $1 - e_2$ for a common e_1, for large A_1 and A_2 (hence large n_1 and n_2).

Proof: For the classical procedure

(3.5) $$e_1 = \mathrm{P}\{V < \hat{C} \mid x_0 \in \pi_1\} \quad \text{and} \quad 1 - e_2 = \mathrm{P}\{V < \hat{C} \mid x_0 \in \pi_2\} ,$$

and similarly for the robust procedure

(3.6) $$e_1 = \mathrm{P}\{V_R < \hat{C}^* \mid x_0 \in \pi_1\} \quad \text{and} \quad 1 - e_2 = \mathrm{P}\{V_R < \hat{C}^* \mid x_0 \in \pi_2\} .$$

Since for large A_1 and A_2 (hence for large n_1 and n_2), the distributions of V and V_R are identical, one only needs to show that $\hat{C}$ is equal to $\hat{C}^*$. As A_1 and A_2 become large, $\hat{C}$ and $\hat{C}^*$ tent to their expected values

(3.7) $$\mathrm{E}(\hat{C}) \simeq (\delta/2) + \frac{1}{2}\left(\frac{1}{n_1} + \frac{1}{n_2}\right) - 1.64 \sqrt{\left\{\delta + \left(\frac{1}{n_1} + \frac{1}{n_2}\right)\right\}}$$

and

(3.8) $$\mathrm{E}(\hat{C}^*) \simeq (\delta/2) + \frac{1}{2}\left(\frac{1}{m_1} + \frac{1}{m_2}\right) - 1.64 \sqrt{\left\{\delta + \left(\frac{1}{m_1} + \frac{1}{m_2}\right)\right\}} .$$

For $n_1 = n_2 = 50$ and $\delta = 2$, for example,

$$\mathrm{E}(\hat{C}) \simeq 1 + 0.02 - 1.64 \sqrt{(2 + 0.04)} = -1.3224 ,$$

and ($q = r/n = 0.1$, $r = 5$, $\beta = 0.8309$, $m_1 = m_2 = 50 - 10 + 8.309 = 48.309$; Appendix B)

$$\mathrm{E}(\hat{C}^*) \simeq 1 + 0.0207 - 1.64 \sqrt{(2 + 0.04140)} = -1.3225 .$$

Since (3.7) and (3.8) are equal for large A_1 and A_2, the result follows.

It is very difficult (if not impossible) to evaluate the values of e_1 and $1 - e_2$ of the V- and V_R-procedures for non-normal populations, especially for small sample sizes n_1 and n_2. We, therefore, resort to Monte Carlo methods. It will suffice to take $n_1 = n_2 = n$; the values for $n_1 \neq n_2$ are exactly similar.

4. Monte Carlo Results

To investigate the robustness of e_1-values, we considered several symmetric distributions and generated n independent random x_1- and x_2-values from the following distributions:

(4.1) (1) Normal; (2) Logistic; Outlier models: (3) $(n-1)\,N(0,1)$ & IN $(0, \sqrt{10})$, (4) $(n-1)\,N(0,1)$ & IN(0, 10), (5) $(n-2)\,N(0,1)$ & 2N(0, 4); Mixture model: (6) 0.90N(0, 1) +0.10 N (0.4) .

These distributions represent a wide class of symmetric distributions prevalent in practice. To create a location shift between the populations π_1 and π_2, a constant d was added to all the x_2-values.

We simulated (from 1000 Monte Carlo runs) the e_1-values of the above V- and V_R-procedures (2.11) and (3.3) and these are given below; the standard errors in these values are well within ± 0.01.

Simulated e_1-values of the V- and V_R-procedures; $n_1=n_2=20$

Distribution d		(1)	(2)	(3)	(4)	(5)	(6)
0.5	V	.057	.055	.035	.011	.020	.037
	V_R	.056	.067	.049	.058	.049	.065
1.5	V	.056	.052	.032	.011	.018	.039
	V_R	.055	.060	.044	.047	.042	.064
3.0	V	.051	.046	.031	.005	.017	.039
	V_R	.042	.052	.038	.036	.037	.060

It is clear that the e_1-values of the V_R-procedure are more stable (robust) from distribution to distribution than the classical V-procedure.

We also simulated the values of $1-e_2$ for all the distributions (4.1): these values are given below:

Simulated values of $1-e_2$ ($e_1 \simeq 0.05$, supposedly); $n_1=n_2=20$

Distribution d		(1)	(2)	(3)	(4)	(5)	(6)
0.5	V	.145	.138	.101	.021	.047	.092
	V_R	.156	.169	.135	.122	.109	.164
1.5	V	.453	.448	.357	.107	.198	.252
	V_R	.458	.484	.440	.406	.381	.395
3.0	V	.918	.916	.858	.378	.647	.653
	V_R	.915	.924	.905	.894	.859	.794

The superiority of the V_R-procedure over the the classical V-procedure is obvious.

Theorem 3: For symmetric long-tailed distributions (kurtosis $\beta_2^* > 3$), the V_R-procedure has higher $1-e_2$ values than the V-procedure, asymptotically (A_1 and A_2 both tend to infinity).

Proof: For the V- and V_R-procedures

$$1-e_2 = \mathrm{P}\{V < \tilde{C} \mid x_0 \in \pi_2\} \quad \text{and} \quad 1-e_2 = \mathrm{P}\{V_R < \tilde{C}^* \mid x_0 \in \pi_2\},$$

respectively. Asymptotically, $\bar{x}_1$ and μ_1 are equal to their expected values $\mathrm{E}(\bar{x}_1) = \mathrm{E}(\hat{\mu}_1) = \mu_1$, $\bar{x}_2$ and μ_2 are equal to their expected values $\mathrm{E}(\bar{x}_2) = \mathrm{E}(\hat{\mu}_2) = \mu_2$, and s is equal to its expected value $\mathrm{E}(s) = \sigma$ and $\hat{\sigma}$ is equal to its expected value $\mathrm{E}(\hat{\sigma}) = k\sigma$ ($k < 1$ for symmetric long-tailed distributions; TIKU, 1980, p. 128; TIKU and KUMRA, 1981). Asymptotically, therefore $\mathrm{E}(V)$ and $\mathrm{Var}\,(V)$ are given by (2.5), but

$$\text{(4.2)} \qquad \mathrm{E}\,(V_R \mid x_0 \in \pi_2) = -(\delta/2)/k^2 \quad \text{and} \quad \mathrm{Var}\,(V_R) = \delta/k^4\,;$$

moreover, $\tilde{C}$ and $\tilde{C}^*$ are equal to their expected values $(\delta/2) - 1.64\sqrt{\delta}$ and $(\delta^*/2) - 1.64\sqrt{\delta^*}$, respectively, $\delta^* = \delta/k^2$.

Define

$$\text{(4.3)} \qquad Z_1 = \{V - \mathrm{E}(V)\}/\sqrt{\{\mathrm{Var}\,(V)\}} \quad \text{and} \quad Z_2 = \{V_R - \mathrm{E}(V_R)\}/\sqrt{\{\mathrm{Var}\,(V_R)\}}\,.$$

Since Z_1 and Z_2 are asymptotically linear in x_0, they have identical distributions. Now, asymptotically,

$$\text{(4.4)} \qquad \mathrm{P}\{V < \tilde{C} \mid x_0 \in \pi_2\} = \mathrm{P}\{Z_1 < \sqrt{\delta} - 1.64\} \quad \text{and} \quad \mathrm{P}\{V_R < \tilde{C}^* \mid x_0 \in \pi_2\} = \mathrm{P}\{Z_2 < \sqrt{\delta} - 1.64k\}\,.$$

Since $0 < k < 1$, the result follows; $k = 1$ if π_1 and π_2 are normal.

Although the V_R-procedure has smaller e_2-values than the V-procedure, but the latter has smaller e_1-values for all the distributions (2) to (6). To compare the V and V_R-procedures, therefore, we adopt a measure of efficiency suggested by several authors (see GESSAMAN and GESSAMAN, 1972, for example), namely, the average $(e_1 + e_2)/2$; smaller values indicating a more efficient classification procedure.

We calculated the values of $(e_1 + e_2)/2$ and found the V_R-procedure considerably more efficient than the V-procedure. For $n_1 = n_2 = 20$ and $d = 3.0$, for example, we have the following values of the average $(e_1 + e_2)/2$; $e_1 \simeq 0.05$ (supposedly):

Distribution	(1)	(2)	(3)	(4)	(5)	(6)
V	.066	.065	.086	.314	.185	.193
V_R	.064	.064	.066	.071	.089	.133

We also simulated the averages for all the distributions (1) to (6) for $n = 30$, 40 and 50 and, as expected, found the V_R-procedure considerably more efficient than the V-procedure. For $n_1=n_2=50$ and $d=2.5$, for example, the average for the V_R and V-procedures are 0.119 and 0.119 for the distribution (1) and 0.122 and 0.271 for the distribution (4), respectively.

5. Classification of a Group of Observations

Suppose that we have a new independent random sample x_{0i}, $i=1, 2, ..., n_0$, which has come from either π_1 or π_2 but we do not know which one. Let $\hat{\mu}_0$ and $\hat{\sigma}_0$ be the MML estimators calculated from the sample x_{0i}, $i=1, 2, ..., n_0$. The robust procedure is to classify the entire x_0-sample in π_1 or π_2 according as

(5.1) $$\bar{V}_R \gtrless C^* ,$$

where

(5.2) $$V_R=\{\hat{\mu}_0-(\hat{\mu}_1+\hat{\mu}_2)/2\}\,(\hat{\mu}_1-\hat{\mu}_2)/\hat{\sigma}^2 ;$$

$\hat{\sigma}^2$ is the pooled MML estimator

(5.3) $$\hat{\sigma}^2=\{(A_1-1)\,\hat{\sigma}_1^2+(A_2-1)\,\hat{\sigma}_2^2+(A_0-1)\,\hat{\sigma}_0^2\}/(A_1+A_2+A_0-3), \quad (A_0=n_0-2r_0) .$$

The constant C^* is so chosen that one of the errors of misclassification (e_1, say) is equal to a pre-specified level (usually, 0.05 or 0.01). If $n_0<5$, then $\hat{\mu}_0$ and $\hat{\sigma}_0^2$ in (5.2) and (5.3) may be replaced by $\bar{x}_0=\sum_{i=1}^{n_0} x_{0i}/n_0$ and $s_0^2=\sum_{i=1}^{n_0} (x_{0i}-\bar{x}_0)^2/(n_0-1)$, respectively.

The classical procedure is to classify the entire x_0-sample in π_1 or π_2 according as

(5.4) $$\bar{V} \gtrless C ;$$

$\bar{V}$ is the statistic $\bar{V}_R$ with $r_1=r_2=r_0=0$, replacing the MML estimators $\hat{\mu}$ and $\hat{\sigma}^2$ by the sample means and variances.

The estimators of C and C^* are obtained as before and are given by ($e_1 \simeq 0.05$, supposedly)

(5.5) $$\tilde{C}=(\tilde{\delta}/2)-1.64\sqrt{(\tilde{\delta}/n_0)} \quad \text{and} \quad \tilde{C}^*=(\tilde{\delta}^*/2)-1.64\sqrt{(\tilde{\delta}^*/m_0)}, \quad (m_0=n_0-2r_0+2r_0\beta_0) ;$$

$\tilde{\delta}$ and $\tilde{\delta}^*$ are given by (2.10) and (3.4), respectively.

We simulated the e_1- and e_2-values of the $\bar{V}$- and $\bar{V}_R$-procedures for all the distributions (1) to (6) and, as expected, found the $\bar{V}_R$-procedure considerably more efficient overall than the $\bar{V}$-procedure. For $n_1=n_2=20$, $n_0=5$ and $d=2.0$, for example, we have the following simulated (based on 1000 Monte Carlo runs) values:

Distribution	e_1		$1-e_2$	
	$\bar{V}$	V	$\bar{V}$	V_R
(1)	.062	.063	.996	.989
(4)	.013	.054	.679	.985
(6)	.062	.067	.866	.950

As n_0 becomes large (and n_1 and n_2 tend to infinity), $\hat{\sigma}$ converges to its expected value $E(\hat{\sigma})$ and $\sqrt{m}\hat{\mu}$, being a linear function of order statistics, is normally distributed; $\bar{V}_R$ and $\bar{V}$ are, therefore, both normally distributed (note that we have assumed that π_1 and π_2 have finite means and variances). In practice, however, n_0 will be small and this result is therefore purely of theoretical interest.

6. Comparison with Nonparametric Procedures

It is of interest to compare the performance of the V- and V_R-procedures with the available nonparametric or distribution-free procedures. Among these, the most prominent are the ones which are based on density estimates.

Let $\hat{f}(z)$ and $\hat{g}(z)$ be consistent estimators of the density functions $f(z)$ and $g(z)$ of the populations π_1 and π_2, respectively. Let z_0 be a new observation that has to be classified. If

$$(6.1) \qquad \hat{f}(z_0)/\hat{g}(z_0) \quad \begin{cases} >1, \text{ classify } z_0 \text{ in } \pi_1 \\ <1, \text{ classify } z_0 \text{ in } \pi_2 \\ =1, \text{ classify } z_0 \text{ arbitrarily in } \pi_1 \text{ or } \pi_2 . \end{cases}$$

The density estimates most commonly used in (6.1) are

(a) The Parzen (1962) kernel estimator

$$(6.2) \qquad f'_m(z) = \frac{1}{mh(m)} \sum_{j=1}^{m} K\left(\frac{z-x_j}{h(m)}\right),$$

where $h(m) = m^{-1/8}$ and $K(w) = \exp(-w^2/2)/\sqrt{(2\pi)}$

(b) The Loftsgaarden-Quesenberry (1965) estimator

$$(6.3) \qquad f'_m(z) = \{(k(m)-1)/m\}/\{2r_{k(m)}\},$$

where $r_{k(m)}$ is the distance from z to the $k(m)$-th closest x_i to z as determined by the Euclidean distance, and the value of $k(m)$ is chosen to be nearer to $m^{1/2}$, as suggested by Loftsgaarden and Quesenberry (1965).

Hudimoto (1964) defined Mann-Whitney type statistics, namely, $\hat{p}_1 = U_{n_1}/n_1$ and $\hat{p}_2 = 1 - U_{n_2}/n_2$, where U_{n_1} is the number of x_{1i}, $i = 1, 2, \ldots, n_1$, less than z_0, and U_{n_2} is the number of x_{2i}, $i = 1, 2, \ldots, n_2$, less than z_0. Hudimoto proposed classifying z_0 in π_1 if and only if $\hat{p}_1 < \hat{p}_2$.

Gupta and Kim (1978) proposed a procedure, similar to Hudimoto's, based on the Parzen kernel density estimator,

$$(6.4) \qquad \hat{\hat{p}}_1 = \frac{1}{n_1} \sum_{i=1}^{n_1} \Phi\left(\frac{z_0 - x_{1i}}{h(n_1)}\right) \quad \text{and} \quad \hat{\hat{p}}_2 = 1 - \frac{1}{n_2} \sum_{i=1}^{n_2} \Phi\left(\frac{z_0 - x_{2i}}{h(n_2)}\right),$$

where $\Phi(.)$ is the cumulative distribution function of a standard normal variate, and $h(m) = m^{-1/8}$; z_0 is classified in π_1 if and only if $\hat{\hat{p}}_1 < \hat{\hat{p}}_2$.

If $S_{n_1}(z)$ and $S_{n_2}(z)$ are the empirical distribution functions of the two samples from π_1 and π_2, respectively, then STOLLER (1954) defined the function

$$(6.5) \qquad \hat{Q}(z) = [n_1 S_{n_1}(z) + n_2 \{1 - S_{n_2}(z)\}]/(n_1 + n_2)$$

and suggested classifying z_0 in π_1 if and only if $z_0 < \eta^*$, where the cut-off point η^* is any value of z that maximizes $\hat{Q}(z)$; for more details, see STOLLER (1954).

DAS GUPTA (1964) studied classification rules based on Wilcoxon statistic and heuristically proposed classifying z_0 in π_1 if and only if $|u - 1/2| < |v - 1/2|$, where $u =$ (number of x_{1i} greater than $z_0)/n_1$ and $v =$ (number of x_{2i} greater than $z_0)/n_2$.

DAS GUPTA (1964) proposed another classification procedure based on Kolmogorov distance; denote by S_i, $(i = 1, 2)$, the sample distribution functions based on the random samples from π_1 and π_2, and denote by d_{0i} the Kolmogorov distance; see DAS GUPTA (1964) for more details. Das Gupta proposed classifying z_0 in π_i, if $d_{0i} = \min \{d_{01}, d_{02}\}$, $(i = 1, 2)$.

Two other interesting procedures are (i) the procedure based on 'statistically equivalent blocks' due to ANDERSON (1966) and GESSAMAN (1970), and (ii) the nearest-neighbour rule of FIX and HODGES (1951), with the distance function $d(x_i, z_0)$ equal to the usual Euclidean distance.

The distributions of all the above procedures are not known, however.

The V- and V_R-procedures equivalent (e_1 not restricted to assume preassigned levels) to the above procedures are

$$(6.6) \qquad V^* \overset{>}{\underset{=}{<}} 0 \quad \text{and} \quad V_R^* \overset{>}{\underset{=}{<}} 0\ ;$$

the expressions for V^* and V_R^* are given by the numerators of (2.3) and (3.2), respectively. For normal populations, the first four moments of V^* and V_R^* can be obtained on the lines of Appendix A.

To investigate the efficiencies of the above classification procedures, we did a Monte Carlo study, following exactly the lines of GESSAMAN and GESSAMAN (1972). However, we excluded the procedure based on 'statistically equivalent blocks' since it requires unrealistically large sample sizes (of the order of 700).

The simulated values (based on 1000 runs) of the error rates $(e_1 + e_2)/2$ of the Parzen P-procedure, the Loftsgaarden-Quesenberry $L - Q$ procedure, the Hudimoto H-procedure, the Gupta-Kim $G - K$ procedure, the Stoller S-procedure, the Das Gupta Wilcoxon W-procedure and Kolmogorov K-procedure, the Fix-Hodges $F - H$ procedure, and the V^*_- and V_R^*-procedures, are given in Table I, for $n_1 = n_2 = 20$ and $d = 2.0$ and 3.0. It is clear that the V_R^*-procedure is superior overall to all others.

We also simulated the error rates for larger sample sizes n_1 and n_2 and found the V_R^*-procedure retaining its superiority. For $n_1 = n_2 = 50$ and $d = 2.5$, for example, we have the following simulated values; the distributions (1) to (5) are given by (4.1):

Distribution	V_R^*	V^*	$F-H$	$L-Q$	P	$G-K$	H	W	K	S
(1)	.104	.105	.113	.110	.109	.108	.111	.125	.119	.120
(2)	.097	.098	.102	.103	.102	.099	.101	.118	.114	.098
(3)	.103	.108	.113	.110	.108	.110	.110	.123	.119	.117
(4)	.104	.110	.113	.110.	.109	.109	.112	.121	.121	.119
(5)	.105	.107	.115	.110	.107	.109	.113	.123	.122	.116

We also simulated the error rates for numerous symmetric short-tailed distributions and skew distributions and here also found the V_R^*-procedure superior overall to all other procedures. For the following distributions

(8) $\chi_6^2/\sqrt{12}$, (9) $\chi_4^2/\sqrt{8}$, (10) Exponential, and (11) Uniform,

for example, we have the following error rates; $n_1 = n_2 = 20$ and $d = 2.5$:

Distribution	V_R^*	V^*	$F-H$	$L-Q$	P	$G-K$	H	W	K	S
(8)	.089	.096	.089	.093	.082	.100	.112	.139	.144	.094
(9)	.062	.064	.078	.076	.073	.069	.090	.099	.108	.084
(10)	.061	.062	.069	.065	.066	.058	.084	.101	.107	.069
(11)**	.233	.233	.248	.248	.230	.238	.237	.255	.270	.237

** Here $d = 0.5$

Note that the computations involved in all the above nonparametric or distribution-free procedures are formidable, but the computations of the V^*- and V_R^*-procedures are straightforward.

Although the V_R^*-procedure has smaller error rates than the V^*-procedure but, asymptotically, the two procedures will have exactly the same error rates; this follows from the following result.

Lemma: Asymptotically, the V^*- and V_R^*-procedures have exactly the same error rates for symmetric populations π_1 and π_2.

Proof: Asymptotically, $\bar{x}_1$ and $\hat{\mu}_1$ are equal to their expected values $E(\bar{x}_1) = E(\hat{\mu}_1) = \mu_1$, and $\bar{x}_2$ and $\hat{\mu}_2$ are equal to their expected values $E(\bar{x}_2) = E(\hat{\mu}_2) = \mu_2$. Both V^* and V_R^* are then identical with the statistic

(6.7) $$Z = \{x_0 - (\mu_1 + \mu_2)/2\}\ (\mu_1 - \mu_2)\ ,$$

and the result follows.

Note that for the normal populations (means known), Z is the likelihood ratio statistic $\log\{f(x;\mu_1,\sigma)/f(x;\mu_2,\sigma)\}$.

Remark: For skew populations, V^* is asymptotically the same as (6.7). Since $E(\hat{\mu}_1 - \hat{\mu}_2)$ is almost equal to $\mu_1 - \mu_2$ and $E\{(\hat{\mu}_1 + \hat{\mu}_2)/2\} = (\mu_1 + \mu_2)/2 + \text{Bias}$, V_R^* is a little different from (6.7); however, the Bias is negligible (TIKU, 1982b, Table I) and, therefore, the V^*- and V_R^*-procedures have asymptotically the same error rates (almost).

The multivariate generalizations of the V_R-procedures are contemplated, on the same lines as TIKU and SINGH (1982). We will report the findings in a later paper.

Of course, a number of estimators of μ, and the matching estimators of σ, are available which could be used in the V_R-procedure, but the MML estimators $\hat{\mu}$ and $\hat{\sigma}$ have some definite advantages; see DUNNETT (1982), TAN (1982) and TIKU (1980)

Table I

Simulated values of the error rates $(e_1+e_2)/2$; $n_1=n_2=20$

Distribution*	d	V_R^*	V^*	$F-H$	$L-Q$	P	$G-K$	H	W	K	S
(1)	2	.150	.151	.157	.167	.160	.157	.157	.184	.186	.168
	3	.054	.055	.062	.062	.059	.060	.072	.102	.098	.078
(2)	2	.147	.147	.155	.154	.150	.147	.149	.172	.171	.162
	3	.070	.071	.072	.069	.071	.069	.076	.089	.099	.078
(3)	2	.153	.153	.155	.167	.161	.157	.153	.187	.191	.171
	3	.055	.060	.057	.064	.058	.061	.073	.102	.102	.077
(4)	2	.155	.180	.156	.169	159	159	152	189	184	169
	3	.056	.072	.057	.064	.060	.061	.077	.104	.095	.077
(5)	2	.155	.154	.163	.165	.163	.164	.158	.187	.190	.171
	3	.057	.060	.057	.064	.059	.065	.070	.099	.092	.083
(6)	2	.190	.196	.198	.202	.200	.186	.193	.229	.231	.209
	3	.122	.124	.124	.121	.125	.123	.130	.163	.159	.125
(7)**	2	.256	.263	.275	.279	.275	.265	.270	.297	.290	.285
	3	.172	.179	.180	.189	.185	.177	.183	.196	.260	.197

* Distributions (1) to (6) are given by (4.1)

** This is Student's t distribution with 4 degrees of freedom

Example 1: Consider the following data which represent the values of $(x-2.0)/(0.1)$, x being the pollution levels (measurements of lead) in the water-samples from two lakes;

Sample from π_1:	−1.4827,	1.2503,	−0.5115,	0.4581,	0.6020,
	−4.2716,	0.6342,	−0.1413,	−0.3753,	1.2790,
	0.9295,	0.5089,	1.1076,	−0.1673,	−0.7866,
	−1.0174,	−0.9106,	0.1018,	0.4125,	1.1147
Sample from π_2:	1.3175,	1.8123,	−0.5443,	2.6796,	2.2655,
	2.6964,	0.7806,	−4.6177,	1.8838,	0.8629,
	2.8592,	0.4699,	−0.4152,	0.1573,	0.6877,
	0.7765,	1.7206,	1.5724,	2.1359,	1.6209.

Observation to be classified is 0.5995 (this has in fact come from π_1).

We have, $\bar{x}_1=-0.0633$, $\bar{x}_2=1.0361$ and $V^*=-0.174$;

$$\hat{\mu}_1=0.1097, \quad \hat{\mu}_2=1.2467 \quad \text{and} \quad V_R^*=0.089 .$$

The classical V^*-procedure wrongly classifies the observation in π_2, whereas the V_R^*-procedure correctly classifies the observation in population π_1.

Example 2: Consider the following data which is similar to that of Example 1;

Sample from π_1:	0.6264,	−0.8991,	−0.4365,	2.7278,	−0.2140,
	0.8573,	−1.2384,	0.3734,	1.6335,	0.7273,
	−0.8753,	−1.3445,	−0.1264,	−0.3433,	1.4616,
	−1.0012	0.0913	0.1148	1.1025,	−1.7202
Sample from π_2:	1.1658,	2.4108,	1.4260,	−0.6028,	1.8013,
	1.0290,	1.4876,	1.2970,	0.4892,	−0.5411,
	−0.3656,	2.2152,	0.1420,	1.7431,	0.4575,
	1.9465,	0.2863,	0.4277,	3.5001,	1.4200.

Observation to be classified is −1.2324 (this has in fact come from π_1).

We have, $\bar{x}_1 = 0.0758$, $\bar{x}_2 = 1.0868$ and $V^* = 1.83$;

$$\hat{\mu}_1 = 0.0306, \quad \hat{\mu}_2 = 1.0372 \quad \text{and} \quad V_R^* = 1.78.$$

Hence, both the procedures correctly classify the observation x_0 in population π_1.

References

ANDERSON, T. W., 1951: Classification by multivariate analysis. Psychometrika **16**, 31–50.

ANDERSON, T. W., 1966: Some nonparametric multivariate procedures based on statistically equivalent blocks. Proceedings of an International Symposium on Multivariate Analysis. New York: Academic Press.

DAS GUPTA, S., 1964: Nonparametric classification rules. Sankhyā **A 26**, 25–30.

DUNNETT, C. W., 1982: Robust multiple comparisons. Commun. Statist.–Theor. Meth., **11** (22), 2611–29. Special Issue on Robust Inference, (Ed. M. L. TIKU).

FIX, E. and HODGES, J. L. Jr., 1951: Discriminatory analysis; Nonparametric discrimination; Consistency properties: USAF School of Aviation Medicine, Report No. 4.

GESSAMAN, M. P., 1970: A consistent nonparametric multivariate density estimator based on statistically equivalent blocks. Ann. Math. Statist. **41**, 1344–6.

GESSAMAN, M. P. and GESSAMAN, P. H., 1972: A comparison of some multivariate discrimination procedures. J. Amer. Statist. Assoc. **67**, 468–72.

GUPTA, A. K. and KIM, B. K., 1978: On a distribution-free discriminant analysis. Biom. J. **20**, 729–36.

HUDIMOTO, H., 1964: On a distribution-free two-way classification. Ann. Inst. Statist. Math. **16**, 247–53.

JOHNSON, N. L., NIXON, E., AMOS, D. E. and PEARSON, E. S., 1963: Tables of percentage points of pearson curves. Biometrika **50**, 459–98.

LACHENBRUCH, P. A. and MICKEY, M. R., 1968: Estimation of error rates in discriminant analysis. Technometrics **10**, 1–11.

LOFTSGAARDEN, D. O. and QUESENBERRY, C. P., 1965: A nonparametric estimate of a multivariate density function. Ann. Math. Statist. **36**, 1049–51.

PARZEN, E., 1962: On estimation of a probability density and mode. Ann. Math. Statist. **33**, 1065–76.

SINHA, B. K. and GIRI, N., 1975: On the distribution of a random matrix. Commun. Statist. **4**, 1057–63.

STOLLER, D. C., 1954: Univariate two population distribution-free discrimination. J. Amer. Statist. Assoc. **49**, 770–5.

TAN, W. Y., 1982: Sampling distributions and robustness of *t*, *F* and variance-ratio in two samples and ANOVA models with respect to departure from normality. Commun. Statist.–Theor. Meth., **11** (22), 2485–2511. Special Issue on Robust Inference, (Ed., M. L. Tiku).

TIKU, M. L., 1967: Estimating the mean and standard deviation from censored normal samples. Biometrika **54**, 155–65.

TIKU, M. L., 1970: Monte Carlo study of some simple estimators in censored normal samples. Biometrika **57**, 207–11.

TIKU, M. L., 1980: Robustness of MML estimators based on censored samples and robust test statistics. J. Statistical Planning and Inference **4**, 123–43.

TIKU, M. L., 1982a: Testing linear contrasts of means in experimental design without assuming normality and homogeneity of variances. Biometrical J. **24**, 613–27.

TIKU, M. L., 1982b: Robust statistics for testing equality of means or variances. Commun. Statist.–Theor. Meth., **11** (22), 2543–58. Special Issue on Robust Inference (Ed., M. L. TIKU).

TIKU, M. L., 1983: MML Estimators and Robust Statistics. A book manuscript, to be published by Marcel Dekker, Inc: New York and Basel.

TIKU, M. L. and STEWART, D., 1977: Estimating and testing group effects from Type I censored normal samples in experimental design. Commun. Statist.–Theor. Meth., **6** (15), 1485–1501.

TIKU, M. L. and KUMRA, S., 1981: Expected values and variances and covariances of order statistics for a family of symmetric distributions (Student's *t*). Statistical Tables in Mathematical Statistics, Vol. 8: American Mathematical Society, Providence (to appear).

TIKU, M. L. and SINGH, M., 1982: Robust statistics for testing mean vectors of multivariate distributions. Commun. Statist.–Theor. Meth., **11** (9), 985–1001.

Appendix A

Write

$$W=\{x_0-(\bar{x}_1+\bar{x}_2)/2\}\,(\bar{x}_1-\bar{x}_2)/s^2\,.$$

When x_0 comes from π_1, the moments of W are given by (under the assumption that π_1 and π_2 are normal)

$$\text{(A.1)}\qquad E(W)=\frac{1}{2}\left\{\delta-\left(\frac{1}{n_1}-\frac{1}{n_2}\right)\right\}\frac{(n_1+n_2-2)}{(n_1+n_2-4)}\,,$$

$$E(W^2)=\left[\frac{1}{4}\,\delta^2+\delta\left(1-\frac{1}{2n_1}+\frac{3}{2n_2}\right)+\frac{1}{n_1}+\frac{1}{n_2}+\frac{1}{n_1n_2}+\frac{3}{4}\left(\frac{1}{n_1}-\frac{1}{n_2}\right)^2\right]$$

$$\text{(A.2)}\qquad \times\frac{(n_1+n_2-2)^2}{(n_1+n_2-4)\,(n_1+n_2-6)}\,,$$

$$E(W^3)=\left[\frac{1}{8}\,\delta^3+\frac{3}{8}\,\delta^2\left(4-\frac{1}{n_1}+\frac{5}{n_2}\right)+\frac{9}{8}\,\delta\left(\frac{8}{n_2}+\frac{1}{n_1^2}+\frac{5}{n_2^2}-\frac{2}{n_1n_2}\right)\right.$$

$$\left.-\frac{9}{8}\left(\frac{1}{n_1}-\frac{1}{n_2}\right)\left\{4\left(\frac{1}{n_1}+\frac{1}{n_2}\right)+\frac{5}{3}\left(\frac{1}{n_1^2}+\frac{1}{n_2^2}\right)+\frac{2}{3n_1n_2}\right\}\right]$$

$$\text{(A.3)}\qquad \times\frac{(n_1+n_2-2)^3}{(n_1+n_2-4)\,(n_1+n_2-6)\,(n_1+n_2-8)}$$

and

$$\mathrm{E}(W^4)=\left[\frac{1}{16}\delta^4+\delta^3\left(\frac{3}{2}-\frac{1}{4n_1}+\frac{7}{4n_2}\right)+\delta^2\left(3-\frac{3}{2n_1}+\frac{45}{2n_2}+\frac{9}{8n_1^2}+\frac{105}{8n_2^2}\right.\right.$$
$$\left.-\frac{15}{4n_1n_2}\right)+\delta\left(\frac{18}{n_1}+\frac{18}{n_2}-\frac{9}{n_1n_2}-\frac{9}{2n_2^2}+\frac{135}{2n_2^2}-\frac{15}{4n_1^3}+\frac{27}{4n_1^2n_2}\right.$$
$$\left.-\frac{45}{4n_1n_2^2}+\frac{105}{4n_2^3}\right)+9\left(\frac{1}{n_1}+\frac{1}{n_2}\right)^2+\frac{9}{2}\left\{5\left(\frac{1}{n_1^3}+\frac{1}{n_2^3}\right)-\frac{1}{n_1n_2}\right.$$
$$\left.\left.\times\left(\frac{1}{n_1}+\frac{1}{n_2}\right)\right\}+\frac{15}{4}\left\{\frac{7}{4}\left(\frac{1}{n_1^4}+\frac{1}{n_2^4}\right)-\frac{1}{n_1n_2}\left(\frac{1}{n_1^2}+\frac{1}{n_2^2}\right)+\frac{9}{10n_1^2n_2^2}\right\}\right]$$

(A.4) $$\times\frac{(n_1+n_2-2)^4}{(n_1+n_2-4)(n_1+n_2-6)(n_1+n_2-8)(n_1+n_2-10)},$$

where $\delta=(\mu_1-\mu_2)^2/\sigma^2$.

When x_0 comes from π_2, the moments $(-1)^j\mathrm{E}(W^j)$, $j=1, 2, 3, 4$, are given by the above expressions with n_1 and n_2 interchanged.

Using the expressions (A.1) to (A.4), we calculated values of the mean, standard deviation, coefficients of skewness $\beta_1^*=\mu_3^2/\mu_2^3$ and kurtosis $\beta_2^*=\mu_4/\mu_2^2$ of the variable $(W-\delta/2)/\sqrt{\delta}$ $(x_0\in\pi_1)$; some of these values are given below:

	Mean	SD	Skewness	Kurtosis
		$n_1=n_2=25$		
$\delta=$ 9.0	0.0652	1.1426	0.2025	3.9309
16.0	0.0870	1.1780	0.2809	3.9676
		$n_1=20$, $n_2=30$		
$\delta=$ 9.0	0.0623	1.1393	0.1936	3.9251
		$n_1=n_2=40$		
$\delta=$ 9.0	0.0395	1.0858	0.0784	3.4912
		$n_1=n_2=50$		
$\delta=$ 9.0	0.0312	1.0678	0.0484	3.3712
		$n_1=n_2=100$		
$\delta=$ 9.0	0.0153	1.0331	0.0121	3.1653

Appendix B

Let

(B.1) $X_{r+1}, \quad X_{r+2}, \ldots, X_{n-r}$

be a Type II censored sample, obtained by arranging n random observations $x_1, x_2, \ldots, x_n$ in ascending order of magnitude and censoring the r smallest and largest observations. If the underlying distribution is normal $N(\mu, \sigma)$, then TIKU's

(1967) MML estimators (defined formally by TIKU and STEWART, 1977) of μ and σ are given by

(B.2) $$\hat{\mu}=\left\{\sum_{i=r+1}^{n-r} X_i+r\beta\,(X_{r+1}+X_{n-r})\right\}/m$$

and

(B.3) $$\hat{\sigma}=\{B+\sqrt{(B^2+4AC)}\}/2\sqrt{\{A\,(A-1)\}}\,,$$

where

$$m=n-2r+2r\beta,\qquad A=n-2r,\qquad B=r\alpha\,(X_{n-r}-X_{r+1})\,,$$

and

(B.4) $$C=\sum_{i=r+1}^{n-r} X_i^2+r\beta\,(X_{r+1}^2+X_{n-r}^2)-m\hat{\mu}^2\,;$$

the coefficients α and β are given by TIKU (1967, eq. 6). For $n>10$, however, α and β are obtained from the following simpler equations (TIKU, 1970)

(B.5) $$\beta=-f(t)\,\{t-f(t)/q\}/q \quad\text{and}\quad \alpha=\{f(t)/q\}-\beta t\,;$$

$q=r/n$, $F(t)=\int_{-\infty}^{t} f(z)\,dz=1-q$ and $f(z)=\{1/\sqrt{(2\pi)}\}\exp\,(-z^2/2)$, $-\infty<z<\infty$. Note that $0<\alpha<1$ and $0<\beta<1$.

For distributions most prevalent in practice, that is, distributions with finite mean and variance, r is chosen to be the integer value $r=[0.5+0.1n]$; see TIKU (1980, 1982).

For symmetric populations of the type $(1/\sigma)\,f\,((x-\mu)/\sigma)$, $E(\hat{\mu})=\mu$ always, and $\hat{\mu}$ and $\hat{\sigma}$ are remarkably efficient (jointly); see TIKU (1980).

To compute $\hat{\mu}$ and $\hat{\sigma}$, all the observations in a sample need not be in fact ordered; one has to locate only the r smallest and largest observations.

For $r=0$, $\hat{\mu}$ and $\hat{\sigma}^2$ reduce to the sample mean and variance $\bar{x}$ and s^2, respectively.

The estimator $\hat{\mu}$ (and $\hat{\sigma}$) satisfies the following efficiency criterion introduced by TIKU (1983):

Let $f(\ldots)$ be a family of distributions which consists of a normal and numerous plausible alternatives to normal. An unbiassed estimator $\hat{\mu}_1$ of the population mean μ is said to be more efficient overall than another unbiassed estimator $\hat{\mu}_2$ if

(B.6) $$\{\sum_f \operatorname{Var}(\hat{\mu}_1)<\sum_f \operatorname{Var}(\hat{\mu}_2)\}_{f\in f(\ldots)}$$

and

$\operatorname{Var}(\hat{\mu}_1)$ is never greater than $\operatorname{Var}(\hat{\mu}_2)$ by more than e for any $f\in f(\ldots)$, say $e=0.01$.

In case of biassed estimators, the variance is replaced by the mean square error. For example, $\hat{\mu}$ and $\hat{\sigma}$ are more efficient overall than $\bar{x}$ and s, respectively, over a family that includes normal and numerous plausible alternatives to normal (TIKU, 1980).

The above efficiency criterion is motivated by the fact that in practice the underlying distribution will hardly ever be exactly normal, in fact will hardly ever be known exactly. It is, therefore, imperative to assume that the underlying distribution is one of a reasonably wide class of distributions, which one exactly is not known.

Manuscript received June 23, 1983

Authors' address:

M. L. Tiku/c.c.
McMaster University
Dept. of Mathematical Sciences
1280 Main Street West
Hamilton, Ontario
Canada L8S 4K1

Biom. J. 27 (1985) 2, 139–148

Robustness of Sequential Tests for Means

D. RASCH

Department of Biostatistics
Research Centre of Animal Production Dummerstorf/Rostock
Academy of Agricultural Sciences of the GDR

Abstract

Three sequential tests of the null hypothesis, $\mu=\mu_0$, versus an alternative of the form $(\mu-\mu_0)^2=\sigma^2d^2$ are compared for different forms of violation of the underlying normal assumption. 10000 samples were simulated for $d=0.6$; 1.0 and 1.6 ($\sigma^2=1$), four (α, β)-combinations and seven alternative distributions. The results show that for small d-values one test is robust for α and another for β. For large d all tests can be used.

Key words and Phrases: normal distribution, truncated normal distribution, comparison of tests, power transformation, violation of distributional assumptions, Monte-Carlo simulation

1. Introduction

The aim of the paper is to present the results of Monte-Carlo-investigations into the robustness of three sequential tests for means of normal variates if the normal assumption is violated. The sequence $\{\underline{y}_1, \underline{y}_2, \ldots\}$ with identically and independently distributed components with mean μ and variance σ^2 $(0<\sigma^2<\infty)$ shall be used to test the hypothesis

$$H_0: \mu=\mu_0$$

against

$$H_A: (\mu-\mu_0)^2=\sigma^2 d^2 \quad (d>0).$$

We consider three tests developed for normally distributed $\underline{y}_i$ by WALD ($\underline{T}_1(n)$, 1947), BARTLETT ($\underline{T}_2(n)$, 1946) and MANN ($\underline{T}_3(n)$, 1980) respectively. The test statistics $\underline{T}_j(n)$ $(j=1, 2, 3)$ are

$$(1) \qquad \underline{T}_1(n)=-\frac{nd^2}{2}+\ln H\left(\frac{n}{2}; \frac{1}{2}; \frac{nd^2\underline{t}_n^2}{2(n-1-\underline{t}_n^2)}\right),$$

$$(2) \qquad \underline{T}_2(n)=-\frac{n-1}{2}\left\{\ln\frac{n-1}{n}\underline{s}_n^2-\ln\left[\frac{n-1}{n}\underline{s}_n^2+(\underline{\bar{y}}-d)^2\right]\right\},$$

$$(3) \qquad \underline{T}_3(n)=\frac{1}{2}\left[\sum_{i=1}^{n}\underline{y}_i^2\right]^{\frac{n-1}{2}}\left\{\left[\sum_{i=1}^{n}(\underline{y}_i+d)^2\right]^{-\frac{n-1}{2}}+\left[\sum_{i=1}^{n}(\underline{y}_i-d)^2\right]^{-\frac{n-1}{2}}\right\}.$$

In (1) to (3) we used

$$\underline{t}_n^2 = \frac{n}{\underline{s}_n^2} (\bar{\underline{y}}_n - \mu_0)^2 ,$$

$$\bar{\underline{y}}_n = \frac{1}{n} \sum_{i=1}^{n} \underline{y}_i ,$$

$$\underline{s}_n^2 = \frac{1}{n-1} \sum_{i=1}^{n} (\underline{y}_i - \bar{\underline{y}})^2$$

and the confluent hypergeometric function $H(a, b, c)$. Usually the decision rule used for the sequential tests is:
Accept H_0 if

$$T_j(n) < b = \ln B \quad (j = 1, 2)$$

or if $T_3(n) < B$.

Reject H_0 if

$$T_j(n) > a = \ln A \quad (j = 1, 2)$$

or if $T_3(n) > A$.

Take a further observation $\underline{y}_{n+1}$, if

$$b \leqq T_j(n) \leqq a \qquad (j = 1, 2)$$

or if $B \leqq T_3(n) \leqq A$

with $A = \frac{1-\beta}{\alpha}$ and $B = \frac{\beta}{1-\alpha}$.

Using this rule, test T_1 has approximately the strength (α, β).

We investigated the robustness of the three tests under different alternative distributions.

Following RASCH and HERRENDÖRFER (1981) we say a test is ε-robust, if the real error probability p_R does not exceed the nominal one by more than ε, that is

$$p_R \leqq p_N (1 + \varepsilon)$$

$(p_R = \alpha_R$ or $p_R = \beta_R)$.

2. The simulation experiment

We considered two systems of alternative distributions so that the simulation experiment consists of two parts. Without loss of generality we put $\mu_0 = 0$ and $\sigma^2 = 1$.

Part I

We investigated the behaviour of the three tests under the normal assumption and for truncated normal distributions. Truncated distributions are characteristic of many investigations in animal and plant husbandry if selection intensity is high.

In the normal case and for one truncated distribution we determined also some points of the OC-curve. Simulation experiments are analogous to experiments in science and have to be planned in the same way. Using the precision requirements used by RASCH and HERRENDÖRFER (1981) we have to generate about 10000 samples.

10000 samples were generated for the normal distribution and one truncated distribution for each of the 264 possible combinations of

$$(\alpha, \beta) = (0.05;\ 0.10)\ (0.05;\ 0.20)\ (0.10;\ 0.20)\ (0.10;\ 0.50)$$
$$d = 0.6;\ 1.0;\ 1.6$$
$$T(n) = T_1(n),\ T_2(n);\ T_3(n)$$
$$\mu = 0\ (0.2)\ d + 0.2$$

and for the other truncated distributions and the distributions of part II only for $\mu = 0$ and $\mu = d$ (72 combinations per distribution). (See Table 1)

If we truncate an $N(\mu^*, \sigma^2)$-distribution in the points x_u and x_0 we have the standardized truncation points

(4) $$u = \frac{x_u - \mu^*}{\sigma} \qquad v = \frac{x_0 - \mu^*}{\sigma}.$$

The truncated distribution is the distribution of a variate $\underline{y}$ with

(5) $$E(\underline{y}) = \mu^* - c_1\sigma,$$

(6) $$V(\underline{y}) = \sigma^2 (1 - c_2 - c_1^2),$$

skewness

(7) $$\gamma_1(\underline{y}) = \frac{c_1 - 2c_1^3 - 3c_1c_2 - c_3}{(1 - c_2 - c_1^2)^{3/2}},$$

Table 1

Distributions used in the simulation experiment and their parameters (u, v standardized truncation points, b, c, d parameters of the power transformation)
No. 1, 2 used in part I, the others in part II

No of distribution	Truncation points u	v	γ_1	γ_2
1	$-\infty$	∞	0	0
2	0.5	3.0	1.0057	0.5915
3	−1.5	3.0	0.3480	−0.3488
4	2.85	4.71	1.505	3.75

	Parameters of power transformation b	c	d		
5	0.748020807992	0	0.077872716101	0	3.75
6	0.63044672784	0	0.11069674204	0	7.00
7	0.953076897706	0.163194276264	0.006597369744	1.00	1.50
8	0.865886203523	0.221027621012	0.027220699158	1.50	3.75

Table 2

Percentage $10^2 f_\alpha$ of false rejection of H_0 for 8 distributions and test 1

			Number of distribution in table 1							
d	$10^2\alpha$	$10^2\beta$	1	2	3	4	5	6	7	8
0,6	5	10	4.19	7.52	4.74	10.38	3.16	2.64	6.12	7.67
		20	4.34	7.44	4.93	10.41	3.23	2.95	6.17	7.57
	10	20	8.85	12.99	9.33	15.91	6.94	6.46	10.52	12.63
		50	9.94	13.79	10.27	15.94	8.30	8.06	11.65	13.70
1.0	5	10	3.90	8.49	4.47	11.06	2.76	2.13	6.40	7.93
		20	4.15	8.67	4.70	11.14	3.03	2.42	6.36	8.06
	10	20	7.90	13.33	9.00	16.31	6.60	5.79	10.65	12.90
		50	10.39	14.64	11.20	17.94	8.78	7.77	12.60	14.69
1.6	5	10	3.87	7.77	4.26	10.76	2.32	2.24	6.04	7.73
		20	4.27	8.01	4.58	11.01	2.60	2.43	6.43	8.14
	10	20	7.68	12.47	8.51	15.58	5.58	5.13	10.51	12.22
		50	11.82	16.33	12.72	19.55	9.70	8.55	14.35	16.06

Table 3

Percentage $10^2 f_\beta$ of false acception of H_0 for 8 distributions and test 1

			Number of distribution in table 1							
d	$10^2\alpha$	$10^2\beta$	1	2	3	4	5	6	7	8
0.6	5	10	7.54	3.26	6.09	1.68	8.00	7.86	4.14	2.37
		20	15.54	8.95	13.78	5.05	14.64	14.76	10.46	7.24
	10	20	15.72	9.43	14.20	5.45	14.34	14.15	10.85	7.66
		50	39.90	34.30	40.79	27.91	34.35	31.97	36.22	30.56
1.0	5	10	5.95	0.35	3.42	0	7.16	7.22	1.22	0.34
		20	11.77	1.52	8.01	0.06	11.90	11.73	3.73	1.27
	10	20	11.58	9.68	8.30	0.07	11.52	11.09	3.94	1.36
		50	30.13	13.13	28.59	0.21	24.25	22.50	18.93	11.23
1.6	5	10	4.30	0	0.61	0	6.83	6.42	0.02	0.01
		20	7.59	0	1.78	0	9.53	9.27	0.12	0.04
	10	20	7.60	0	2.01	0	9.38	8.87	0.13	0.04
		50	15.01	0.17	10.58	0.08	13 98	12.64	2.37	0.44

and kurtosis

(8) $$\gamma_2(y) = \frac{4c_1^2 - 6c_1^4 + 3c_2 - 12c_1^2 c_2 - c_4 - 4c_1 c_3 - 3c_2^2}{(1 - c_2 - c_1^2)^2}.$$

In (5) to (8) we used the abbreviation

(9) $$c_i = \frac{1}{\Delta}\left[v^{i-1}\varphi(v) - u^{i-1}\varphi(u)\right], \qquad (i = 1, 2, 3, 4)$$

with $\Delta = \Phi(v) - \Phi(u)$ and the density φ and the distribution function Φ of the standard normal distribution. To obtain pseudo-random number from a truncated distribution with mean μ, variance 1, skewness γ_1 and kurtosis γ_2 we first generate in (0.1) uniformly distributed pseudo-random numbers (p.r.n.) P and transform them by

(10) $$P' = P\Delta + \Phi(u)$$

Table 4

Average sample number $\bar{n}$ for H_0 (upper entry) and H_A (lower entry) for 8 distributions and test 1

d	$10^2\alpha$	$10^2\beta$	Number of distributions in table 1 1	2	3	4	5	6	7	8
0.6	5	10	20.4	19.7	20.2	19.5	20.9	21.2	20.0	19.9
			22.2	23.9	23.2	23.6	20.5	19.8	23.6	23.5
	5	20	15.1	14.7	15.0	14.6	15.6	16.0	15.0	14.8
			19.0	21.6	20.3	22.0	17.9	17.2	21.1	21.5
	10	20	14.0	13.5	13.9	13.4	14.6	15.0	13.9	13.6
			15.3	17.5	16.2	17.7	14.3	13.7	16.9	17.2
	10	50	7.0	6.8	6.8	6.7	7.3	7.6	6.9	6.9
			8.7	10.4	9.2	11.3	8.4	8.3	10.0	10.6
1.0	5	10	8.3	8.1	8.3	8.1	8.6	8.8	8.3	8.3
			10.4	11.0	10.9	10.3	9.5	9.1	10.7	10.4
	5	20	6.7	6.5	6.6	6.5	6.9	7.0	6.6	6.6
			9.3	10.4	10.0	9.1	8.6	8.2	10.1	10.0
	10	20	6.3	6.1	6.2	6.1	6.5	6.7	6.2	6.2
			7.6	8.5	8.1	7.9	7.0	6.7	8.2	8.0
	10	50	3.7	3.6	3.7	3.7	3.9	4.0	3.7	3.7
			4.9	6.2	5.2	6.5	4.8	4.7	5.7	6.0
1.6	5	10	4.2	4.2	4.2	4.2	4.4	4.4	4.3	4.3
			6.2	6.1	6.5	5.8	5.7	5.5	6.2	5.9
	5	20	3.7	3.6	3.7	3.7	3.8	3.8	3.7	3.7
			5.8	5.9	6.2	5.6	5.3	5.1	6.0	5.7
	10	20	3.5	3.7	3.5	3.5	3.6	3.7	3.5	3.5
			4.8	4.8	5.2	4.6	4.5	4.4	5.0	4.7
	10	50	2.9	2.8	2.9	2.8	2.9	2.9	2.8	2.8
			3.7	4.0	4.0	3.9	3.5	3.4	4.1	3.9

Table 5

Percentage $10^2 f_\alpha$ of false rejection of H_0 for 8 distributions and test 2

d	$10^2\alpha$	$10^2\beta$	Number of distribution in table 1 1	2	3	4	5	6	7	8
0.6	5	10	4.25	1.75	3.44	0.98	3.26	2.51	1.62	1.17
		20	4.64	1.69	3.59	1.11	3.33	2.68	1.78	1.36
	10	20	8.92	3.95	7.11	2.62	6.78	5.94	4.25	3.26
		50	11.15	5.46	9.32	3.96	9.06	8.03	6.26	4.41
1.0	5	10	2.61	1.14	2.00	0.61	1.26	0.77	1.14	0.80
		20	2.65	1.20	2.28	0.61	1.39	0.84	1.31	0.88
	10	20	5.21	2.73	4.92	1.91	3.01	2.12	3.14	1.94
		50	7.21	4.20	7.00	3.18	4.36	3.34	4.50	2.99
1.6	5	10	0.51	0.42	0.67	0.25	0.21	0.18	0.37	0.28
		20	0.58	0.45	0.71	0.26	0.25	0.19	0.42	0.29
	10	20	1.37	1.08	1.42	0.58	0.78	0.42	1.01	0.77
		50	2.37	2.00	2.50	1.29	1.31	0.88	1.84	1.36

into in $(\Phi(u), \Phi(v))$ uniformly distributed p.r.n. Applying a transformation from uniform to normal variates at P', adding c_1, dividing then by $\sqrt{1-c_2-c_1^2}$ and adding μ leads to p.r.n. with the given expectation μ, variance equal to one and the

Table 6

Percentage $10^2 f_\beta$ of false acceptance of H_0 for 8 distributions and test 2

d	$10^2\alpha$	$10^2\beta$	Number of distribution in table 1	2	3	4	5	6	7	8
0.6	5	10	8.37	19.18	11.48	22.85	6.85	5.71	15.48	18.38
		20	15.46	27.80	19.45	31.30	13.15	11.53	24.01	27.01
	10	20	15.51	28.15	19.64	31.58	13.11	11.46	24.10	27.30
		50	32.27	44.53	36.28	46.32	28.56	26.52	39.93	42.27
1.0	5	10	4.45	10.37	6.66	9.55	2.78	1.93	8.07	8.07
		20	8.90	16.70	11.56	14.13	5.96	4.85	13.60	12.90
	10	20	9.17	17.11	11.97	14.55	6.10	4.97	13.94	13.29
		50	20.55	26.81	22.86	24.94	16.04	14.44	24.18	24.32
1.6	5	10	11.19	0.19	1.60	0.01	0.54	0.35	0.76	0.13
		20	12.55	0.85	3.21	0.04	1.31	1.03	1.58	0.53
	10	20	12.65	0.94	3.36	0.06	1.33	1.06	1.67	0.61
		50	8.09	5.34	8.77	1.46	5.75	5.03	6.04	3.77

Table 7

Average sample number $\bar{n}$ for H_0 (upper entry) and H_A (lower entry) for 8 distributions and test 2

d	$10^2\alpha$	$10^2\beta$	Number of distribution in table 1	2	3	4	5	6	7	8
0.6	5	10	15.3	15.2	15.5	14.2	14.2	13.6	14.9	14.2
			18.2	17.2	18.2	15.9	16.9	16.2	17.3	16.6
	5	20	11.1	11.1	11.4	10.4	10.4	9.9	10.9	10.4
			15.8	14.6	15.6	13.6	14.7	14.2	14.7	14.1
	10	20	10.1	10.2	10.4	9.7	9.4	9.1	10.1	9.7
			12.0	11.1	11.9	10.4	11.2	10.8	11.3	10.8
	10	50	5.0	5.3	5.2	5.1	4.8	4.7	5.2	5.0
			7.1	6.8	7.1	6.4	6.6	6.4	6.8	6.6
1.0	5	10	7.9	7.8	8.0	7.5	7.3	7.1	7.7	7.4
			9.4	9.1	9.3	8.8	8.8	8.4	9.0	8.8
	5	20	6.0	6.0	6.2	5.8	5.6	5.4	5.9	5.8
			8.6	8.1	8.5	8.1	8.1	7.7	8.1	8.1
	10	20	5.7	5.7	5.8	5.5	5.4	5.2	5.6	5.5
			6.8	6.4	6.7	6.4	6.4	6.2	6.4	6.4
	10	50	3.3	3.3	3.4	3.3	3.2	3.2	3.3	3.3
			4.7	4.7	4.8	4.7	4.5	4.7	4.7	4.6
1.6	5	10	5.0	5.1	5.1	4.9	4.7	4.6	5.0	4.8
			5.9	6.0	6.0	5.8	5.6	5.4	5.9	5.7
	5	20	4.0	4.1	4.1	4.0	3.8	3.6	4.0	3.9
			5.7	4.8	5.7	5.6	5.4	5.2	5.7	5.5
	10	20	3.9	4.0	3.9	3.7	3.7	3.6	3.8	3.8
			4.6	4.7	4.7	4.6	4.4	4.3	4.6	4.5
	10	50	2.6	2.6	2.6	2.6	2.5	2.5	2.6	2.6
			3.7	3.9	3.8	3.9	3.6	3.5	3.8	3.8

given γ_1- and γ_2-values with u and v in (10); c_1 and c_2 are roots of (7) and (8) with the predetermined γ_1- or γ_2-values on the l.h.s.

Part II P.r.n. from $N(\mu, 1)$ were generated for the 72 parameter combinations

Table 8

Percentage $10^2 f_\alpha$ of false rejection of H_0 for 8 distributions and test 3

d	$10^2\alpha$	$10^2\beta$	Number of distribution in table 1 1	2	3	4	5	6	7	8
0.6	5	10	4.80	8.80	5.26	12.60	3.27	2.41	7.03	6.73
		20	4.90	9.18	5.41	12.54	3.33	3.53	7.15	7.02
	10	20	9.17	14.47	9.32	18.17	6.60	5.43	11.70	12.01
		50	10.77	15.82	11.68	19.10	8.60	7.03	13.30	13.15
1.0	5	10	2.61	5.43	3.46	4.41	1.05	0.58	3.22	4.01
		20	2.80	5.74	3.70	4.63	1.12	0.65	3.54	3.23
	10	20	5.50	8.51	6.53	7.69	2.37	1.69	6.11	5.78
		50	7.23	10.28	8.45	9.42	3.64	2.64	8.40	7.89
1.6	5	10	0.43	0.09	0.66	0.11	0.13	0.06	0.34	0.45
		20	0.48	0.11	0.69	0.16	0.18	0.08	0.39	0.63
	10	20	1.12	0.48	1.47	0.39	0.43	0.23	0.73	0.78
		50	1.80	0.93	2.33	0.66	0.79	0.49	1.22	1.02

Table 9

Percentage $10^2 f_\beta$ of false acception of H_0 for 8 distributions and test 3

d	$10^2\alpha$	$10^2\beta$	Number of distribution in table 1 1	2	3	4	5	6	7	8
0.6	5	10	10.36	19.47	12.60	25.36	9.81	8.86	17.46	16.79
		20	18.90	29.87.	21.68	34.84	17.74	16.80	27.21	22.75
	10	20	18.91	30.09	21.74	35.35	17.67	16.57	27.20	27.15
		50	40.44	49.59	43.46	53.64	37.96	36.37	47.72	45.12
1.0	5	10	6.56	12.72	8.72	12.36	3.77	3.02	9.80	10.25
		20	12.44	19.85	14.88	18.69	8.70	7.28	16.19	15.80
	10	20	12.77	20.34	15.31	19.25	8.86	7.36	16.58	16.03
		50	27.84	34.33	30.88	33.71	23.04	21.12	31.45	29.50
1.6	5	10	1.75	0.59	2.14	0.04	0.82	0.61	0.94	0.99
		20	4.19	2.01	4.76	0.34	2.30	1.82	2.48	2.12
	10	20	4.47	2.17	5.05	0.36	2.48	1.95	2.78	2.55
		50	14.01	10.99	13.71	6.58	10.71	9.29	11.74	10.79

mentioned in part I. These p.r.n. were transformed into p.r.n. with given skewness γ_1 and kurtosis γ_2 by the so-called power transformation $z = -c + by + cy^2 + dy^3$ (FLEISHMAN, 1978, NÜRNBERG, 1982).

For both parts we calculated for each μ-value the relative frequency of the 10000 samples in which H_0 was accepted and used this for $\mu = 0$ as an estimate of α_R and for $\mu = d$ as an estimate of $1 - \beta_R$. Further, we calculated the average $\bar{n}$ of the sizes of the 10000 samples and the variance of these sample sizes s_n^2. We used $\bar{n}$ as an estimate of $E(\underline{n})$. We also determined $n_{\min}$ and $n_{\max}$ and the frequency distribution of the n_l ($l = 1, \ldots, 10^5$).

Table 10

Average sample number $\bar{n}$ for H_0 (upper entry) and H_A (lower entry) for 8 distributions and test 3

d	$10^2\alpha$	$10^2\beta$	Number of distribution in table 1 1	2	3	4	5	6	7	8
0.6	5	10	20.4	19.9	20.3	18.6	18.5	17.4	19.4	19.2
			21.8	21.0	22.1	19.2	20.6	19.7	20.9	20.9
	5	20	14.9	14.5	14.8	13.6	13.3	12.5	14.1	14.0
			18.9	17.5	18.8	16.2	17.7	17.0	17.9	17.5
	10	20	13.7	13.3	13.7	12.4	12.5	11.7	13.0	12.8
			15.1	14.1	15.2	13.0	14.2	13.6	14.3	14.3
	10	50	6.7	6.5	6.7	6.2	6.3	6.0	6.5	6.4
			8.9	8.5	8.9	7.9	8.3	7.9	8.6	8.2
1.0	5	10	9.3	9.2	9.3	8.8	8.5	8.0	9.0	8.8
			10.5	10.6	11.0	10.3	10.4	10.0	10.6	10.5
	5	20	7.0	7.0	7.1	6.7	6.5	6.1	6.9	6.7
			9.9	9.4	9.9	9.3	9.4	9.1	9.5	9.4
	10	20	6.6	6.6	6.7	6.4	6.2	5.9	6.5	6.3
			8.1	7.8	8.1	7.6	7.7	7.4	7.8	7.5
	10	50	3.9	3.9	3.9	3.7	3.7	3.6	3.8	3.7
			5.6	5.6	5.7	5.5	5.3	5.2	5.6	5.5
1.6	5	10	5.4	5.4	5.5	5.2	5.1	4.9	5.4	5.4
			6.9	7.0	7.0	6.7	6.6	6.4	6.9	6.8
	5	20	4.3	4.3	4.4	4.1	4.0	3.9	4.2	4.1
			6.6	6.7	6.6	6.7	6.3	6.1	6.6	6.7
	10	20	4.2	4.2	4.2	4.0	3.9	3.8	4.1	4.1
			5.5	5.7	5.6	5.5	5.3	5.2	5.6	5.5
	10	50	2.7	2.7	2.8	2.7	2.7	2.6	2.7	2.5
			4.3	4.5	4.5	4.6	4.2	4.1	5.4	5.3

3. Results

In a recent paper by RASCH (1983) it was shown that the empirical and the theoretical values of the mean, variance, skewness and kurtosis are in a good aggreement. Empirical and theoretical values are also compared in the present experiment for all the generated distributions in table 1, and again the empirical and theoretical values differed only slightly. To save space the results are not included in this paper: readers who wish to see the degree of agreement are referred to Table 1 in RASCH (1983). Because the values of s_n^2, $n_{\min}$ and $n_{\max}$ differed only slightly between the three tests we decided not to include the tables showing these values. The simulation experiments reveal the following tendencies

(i) With four exceptions the empirical α-values are, under normal conditions lower than the nominal ones. Under normal conditions the empirical $(1-\beta)$-values are with the exception of T_3; $(\alpha, \beta) = (0.05, 0.10)$, higher than the nominal ones, so that all tests are in most cases conservative, as could be expected from theory in the case of test 1.

(ii) With respect to α test 2 is robust for all investigated alternative distributions (it is worst in the normal case), tests 1 and 3 are non-robust for small d if γ_1 and γ_2 both differ greatly from zero

(iii) With respect to β test 1 is always robust, test 2 is robust only if the values of d are not too small and test 3 is robust only for $d=1.6$.

(iv) The ASN is low for test 2 if d is small and is also good for test 1 if d is large. Test 3 always needs larger average sample sizes than either test 1 or test 2. For median d-values, the sample sizes do not differ too much between the three tests.

4. Further research

Because the influence of d on the robustness of sequential tests is very important we intend to investigate further d-values and a test with the statistic

$$\underline{T}_4=\underline{T}_1, \quad \text{if} \quad u<0.5, \quad \underline{T}_4=\underline{T}_2, \quad \text{if} \quad u\geqq 0.5$$

$(p_1+p_2=1)$. We will also extend the investigations to include more alternative distributions (u is a uniformly distributed p.r.n.. in (0.1)).

References

Autorenkollektiv, 1980: Robustheit I.
Probleme der angewandten Statistik Heft 4, FZ für Tierproduktion Dummerstorf-Rostock der AdL der DDR

Autorenkollektiv, 1981: Robustheit II.
Probleme der angewandten Statistik Heft 5, FZ für Tierproduktion Dummerstorf-Rostock der AdL der DDR

Autorenkollektiv, 1982: Robustheit III.
Probleme der angewandten Statistik Heft 7, FZ für Tierproduktion Dummerstorf-Rostock der AdL der DDR

BARTLETT, M. S., 1946: The large sample theory of sequential tests. Proc. Comb. Phil. Soc. **42**, 239–244

FLEISHMAN, A. L., 1978: A method for simulating non-normal distributions,Psychometrika **43**, 521–532

LIM, K. and K. Y. FUNG, 1982: Sequential t tests based on M estimators. Statist. Comp. Simul. **14**, 271–281

MANN, J. E., 1980: A simple sequential criterion for testing a normal mean with unknown variance. Commun. Statist.–Theor. Meth. **A 9 (15)**, 1619–1624.

NÜRNBERG, G., 1982: Beiträge zur Versuchsplanung für die Schätzung von Varianzkomponenten und Robustheitsuntersuchungen zum Vergleich zweier Tests, Probleme der angewandten Statistik Heft 6, FZ für Tierproduktion Dummerstorf-Rostock der AdL der DDR

RASCH, D.; HERRENDÖRFER, G., 1981: Robustheit statistischer Verfahren, Rostock. Math. Kolloq. **17**, 87–104

RASCH, D.; TEUSCHER, F., 1982: Das System der gestutzten Normalverteilungen. 7. Sitzungsbericht der Interessengemeinschaft Mathematische Statistik MG DDR, Berlin, 51–60.

RASCH, D.: First results on robustness of the one-sample sequential t-test. Trans. 9th Prague Conference on Inform. Theory and Mathem. Statist. (in press)

WALD, A., 1947: Sequential Analysis. New York: Wiley

Manuscript received Jan. 20, 1983

Author's address:

Prof. Dr. D. RASCH
FZ für Tierproduktion
Abt. Biomathematik und EDV
DDR - 2551 Dummerstorf
Kreis Rostock

Biom. J. 27 (1985) 2, 148

Book Reviews

HAND, D. J.: Discrimination and Classification. John Wiley & Sons, Chichester – Brisbane – New York – Toronto 1981. XII, 218 pp., $ 15.50.

The role of discrimination and classification increased rapidly in the last years mainly as a consequence of the development of computers and the good applicability of these methods. This book discusses different topics of this field. Mainly it will help to use the discrimination and classification methods and especially the author will increase the understanding for these methods. In such a way he discusses the basic ideas of the problems, gives methods and algorithms and illustrates the results on several examples. At the end of each chapter exercises are formulated. The book is appropriate for statisticians and demands the level of undergraduates in statistics.

In the chapter 1 introductional remarks and notations are summarized. In the chapter 2 classification methods are discussed and some distribution–free procedures are given. The third chapter deals with parameterized distributions and maximum likelihood and Bayes methods are discussed. In the chapter 4 linear discriminant functions are introduced and in the fifth discrete distributions are investigated. The chapter 6 deals with the variable selection and the seventh investigates hierarchical cluster methods.

All results are illustrated by many examples and the book is very appropriate for statisticians who work in applications. It contains the basic ideas very clear.

H. Läuter

Biom. J. 27 (1985) 2, 149–157

Evaluating an Explicit Functional Relationship by Fuzzifying the Statement of its Satisfying

H. BANDEMER and S. SCHMERLING

Bergakademie Freiberg, Sektion Mathematik; Martin-Luther-Universität Halle – Wittenberg, Sektion Mathematik

Abstract

Combining the concepts of interval arithmetic and fuzzy theory the observations to identify some explicit functional relationship are assumed to be fuzzy sets given by the threshold sets of their membership function, which have contours suitable for the application of interval arithmetic procedures. The result is a fuzzy set on the parameter region, which offers possibilities for reasoning on the validity of the relationship. The proceeding is illustrated by an example.

Key words: fuzzy approach; functional relationships

1. Introduction

When investigating the dependence of some response variable y on several regressors, say $\underline{x}=(x_1, \ldots, x_k)'$, we use to perform experiments, the results of which we mark by points $(x_{1i}, \ldots, x_{ki}, y_i)=(\underline{x}_i, y_i)$; $i=1, \ldots, n$; in some region $Z=B\times R^1\subset R^{k+1}$, with $\underline{x}_i\in B$ and $y\in R^1$.

Being quite aware of the fact that this is an insufficient proceeding to identify some explicit functional relationship, say

$$(1.1) \qquad y=f(\underline{x}); \qquad \underline{x}\in B$$

governing (approximately) the mutual dependence of y and $\underline{x}$, we specify a certain model, on which we base our reasoning.

Usually we choose a model from mathematical statistics saying that the functional relationship in question is superimposed by some random noise and try to come to statements on the relationship using a certain approximation principle working as an estimator. In some cases we can assume the values of the regressors $\underline{x}$ to be known exactly and random errors to occur only with the response y. Even that, however, seems to be a rough simplification. As a rule, we must allow both, responses and regressors, to be subject to inaccuracy. For such situations the common procedures of mathematical statistics either demand for crucial assumptions, e.g. as with respect to the joint distribution of $(\underline{x}, y)$ or supply only poor statements concerning the actual properties of the estimation obtained.

One way to overcome these difficulties was suggested by SCHMERLING (1980). He starts in an interval arithmetic style assuming that the "observation points" are given by certain subsets $Q_j \subset Z : j=1, ..., n$; each containing at least one point of the graph (1.1). Having the relationship parametrized by $\underline{\vartheta} \in \Theta = R^r$ he evaluates some estimation region, say $G \subset R^r$, containing all $\underline{\vartheta}$ of $y=f(\underline{x}, \underline{\vartheta})$, for which the graph of the relationship hits each of the given observations Q_j. This approach was generalized in BANDEMER/SCHMERLING (1983) to the case that there are given certain probabilities c_j of the events: Q_j contains at least one point of the true but unknown relationship, say $y=f(\underline{x}, \underline{\vartheta}^{(0)})$.

From this standpoint it seems only a step to come to an approach, which uses fuzzy sets. Some proposals in this direction were given by BANDEMER (1983).

He considers the observations as fuzzy sets M_i given by their membership functions

(1.2) $\mu_i(\underline{x}, y) \mid Z \to [0, 1]$

from which he calculates fuzzy sets on the set of all possible parameters $\underline{\vartheta} \in \Theta$ taking either the expectation value with respect to a prior probability measure P on $[B, \underline{\mathfrak{L}}]$ or the fuzzy expectation with respect to a prior fuzzy measure g on $[B, \underline{\mathfrak{L}}]$, after having collected the sets M_i in the usual manner. However, there can occur two difficulties with this approach: (i) the specification of P or g, resp. and (ii) the fact that only a subset of all the given observations operates, when calculating the regions in Θ.

In the following we will suggest another approach combining both modelling proposals assuming that the observations are fuzzy and demanding that the relationship in question hits all these observations in a fuzzy meaning, which we have to define.

2. Fuzzy Statements With Respect to a Relationship

In non-fuzzy theory we deal with statements of the following kind:

Statement S_0: A given functional relationship (1.1) is satisfied in a given point

$$(\underline{x}_0, y_0).$$

This statement is true, if $y_0 = f(\underline{x}_0)$, and false, if $y_0 \neq f(\underline{x}_0)$.

Statement S: A given functional relationship (1.1) is satisfied in a given set

$$M \subset Z.$$

This statement is true, if there exists some $(\underline{x}_0, y_0) \in M$ with $y_0 = f(\underline{x}_0)$, and it is false otherwise.

To come to a fuzzification of S we define a fuzzy observation. We consider a fuzzy set M, given by its membership function

(2.1) $\mu_M \mid Z \to [0, 1]$,

and introduce its threshold sets

$$\forall \alpha \in [0, 1] : C_\alpha(M) := \{(\underline{x}, y) \in Z : \mu_M(\underline{x}, y) \geqq \alpha\} \tag{2.2}$$

Definition 2.1: The fuzzy set M is called *fuzzy observation*, if for all $\alpha \in (0, 1]$ the sets $C_\alpha(M)$ are bounded, simply connected and closed.

The function μ indicates to what extent each point $\underline{x}$ of M should be considered as a point of the observation. To get an imagination of that fuzzy approach we may interpret μ as a greytone picture taking $\mu = 0$ for white and $\mu = 1$ for black and the fuzzy observation becomes a "dot" of finite area showing a greytone structure. The threshold sets are then the boundaries of the subareas with shadow intensity of at least α, resp.

Now we can fuzzify Statement S.

Definition 2.2: The value $\mu_s(f : M)$ is called *fuzzy grade of validity of S in M if*

$$\mu_s(f; M) = \sup_{\underline{x} \in B} \mu_M(\underline{x}, f(\underline{x})) \tag{2.3}$$

In our interpretation the fuzzy grade of validity of S in M is given by the "darkest point" of M that the graph $(\underline{x}, f(\underline{x}))$ of the relationship hits.

The connection with the case of S with a non-fuzzy set M is supplied by the following lemma, the assumptions of which can be weakened, obviously.

Lemma 2.1: If B is compact, μ_M continuous on Z, and f continuous on B then

$$\mu_s(f; M) = 1 \Leftrightarrow \exists\ \underline{x} \in B : \mu_M(\underline{x}, f(\underline{x})) = 1 \tag{2.4}$$

For the proof it suffices to say that under the given assumptions the supremum in (2.3) is a maximum.

Lemma 2.1 states that under the given conditions we have the value 1 of the grade of validity if and only if there exists at least on point in M where the relationship is satisfied (i.e. $\mu_M(\underline{x}, f(\underline{x})) = 1$), but this means that Statement S is true.

Let be given n fuzzy observations M_i with

$$\mu_{M_i} \mid Z \to [0, 1] \tag{2.5}$$

then we have n fuzzy grades of validity of S in M_i

$$\mu_s(f; M_i) = \sup_{\underline{x} \in B} \mu_{M_i}(\underline{x}, f(\underline{x})) \tag{2.6}$$

Now we turn to the statement, generalizing S_0 in another direction.

Statement S_n: A given functional relationship (1.1) is satisfied in each of n given distinct points, say

$$(\underline{x}_i, y_i); \qquad i = 1, \ldots, n .$$

This statement is true, if for all $i = 1, \ldots, n : y_i = f(\underline{x}_i)$, and false, if there exists some $i \in \{1, \ldots, n\}$ with $y_i \neq f(\underline{x}_i)$. The appropriate fuzzification of S_n demands for "satisfying S in each fuzzy observation":

$$(S_n \text{ in } \{M_1, M_2, \ldots, M_n\}) = \bigwedge_i (S \text{ in } M_i) \tag{2.7}$$

Hence we obtain the fuzzy grade of validity from (2.6) by

$$\mu_{S_n}(f;\ M_1, \ldots, M_n) = \bigwedge_i \mu_S(f;\ M_i) = \min_i \sup_{\underline{x} \in B} \mu_{M_i}(\underline{x}, f(\underline{x})) \tag{2.8}$$

3. Fuzzy Statements and Parametrized Functional Relationships

Now we turn to functional relationships in their parametrized form, i.e. we are given a family of functions

$$\{(\underline{x}, y) \in Z : y = f(\underline{x}, \underline{\vartheta})\}_{\underline{\vartheta} \in \Theta} . \tag{3.1}$$

The relationship is specified by the value $\underline{\vartheta}^{(0)} \in \Theta \subseteqq R^r$. In application the components of this vector are certain parameters, e.g. material or state constants characterizing the problem under consideration. In this context $\underline{\vartheta}^{(0)}$ is called the *true* parameter value. In general it is unknown and must be investigated by means of observations.

We assume these observations to be fuzzy observations M_i in the sense of Definition 2.1. and consider the fuzzy grade of validity (as defined by (2.3)) for every $\underline{\vartheta} \in \Theta$ by

$$\mu_s(f, \underline{\vartheta};\ M_i) = \sup_{\underline{x} \in B} \mu_{M_i}(\underline{x}, f(\underline{x}, \underline{\vartheta})) . \tag{3.2}$$

In the Appendix we will show (Lemma A) that the threshold set of $\mu_{s_n}(f;\ M_1, \ldots, M_n)$ for some $\alpha \in (0, 1]$ is given uniquely by the points $(\underline{x}, y) \in Z$ in which the graph of the functional relationship hits all the n threshold sets $C_\alpha(M_i)$, $i = 1, \ldots, n$ belonging to that α.

With regard to Lemma A it makes sense to introduce, for every $C_\alpha(M_i)$, the following set:

$$G_i(\alpha) = \{\underline{\vartheta} \in \Theta :\ \exists (\underline{x}, y) \in C_\alpha(M_i) : y = f(\underline{x}, \underline{\vartheta})\} \tag{3.3}$$

i.e. the set of all parameter values $\underline{\vartheta} \in \Theta$, for which the graph of the functional relationship hits the α-threshold set of the fuzzy observation:

$$\{(\underline{x}, y) \in Z : y = f(\underline{x}, \underline{\vartheta})\} \cap C_\alpha(M_i) \neq \emptyset . \tag{3.4}$$

Hence we deduce from Lemma A

Corollary 3.1: Under the assumptions of Lemma 2.1 it holds for all $\alpha \in (0, 1]$

$$\mu_s(f, \underline{\vartheta};\ M_i) \geqq \alpha \Leftrightarrow \underline{\vartheta} \in G_i(\alpha) \tag{3.5}$$

Turning over to n fuzzy observations we define

$$\mu_{s_n}(f, \underline{\vartheta};\ M_1, \ldots, M_n) = \min_i \mu_s(f;\ \underline{\vartheta}, M_i) \tag{3.6}$$

and obtain analogously

Corollary 3.2: Under the assumptions of Lemma 2.1 it holds for all $\alpha \in (0, 1]$

$$\mu_{s_n}(f, \underline{\vartheta};\ M_1, \ldots, M_n) \geqq \alpha \Leftrightarrow \forall i \in \{1, \ldots, n\} : \underline{\vartheta} \in G_i(\alpha) \Leftrightarrow \underline{\vartheta} \in \bigcap_i G_i(\alpha) . \tag{3.7}$$

In dependence on $\underline{\vartheta}$ the function

(3.8) $$\mu_D(.) := \mu_{s_n}(f, \underline{\vartheta}; M_1, \ldots, M_n) : \Theta \to [0, 1]$$

defines a fuzzy set D on Θ, which is a transformation (into Θ) of the information on the functional relationship given by the fuzzy observations $M_1, \ldots, M_n$ on Z. As we can see from Lemma A the family of sets

(3.9) $$\{G(\alpha)\}_{\alpha \in (0,1]}; \; G(\alpha) := \bigcap_i G_i(\alpha)$$

is the family of threshold sets $C_\alpha(D)$ of the fuzzy set D:

(3.10) $$\forall \alpha \in (0, 1] : G(\alpha) = C_\alpha(D) .$$

4. Computation of the Threshold Sets in the Parameter Region

The fuzzy set D according to (3.8) represents the information on the relationship (3.1) given by the fuzzy observation (2.5) considered as a whole. Hence evaluating the relationship means computing the membership function μ_D of D. Equivalently we can compute its threshold sets $C_\alpha(D)$ for some $0 < \alpha_1 < \alpha_2 < \ldots < \alpha_T \leq 1$ thus giving a comprehensible impression of its contour. The procedure of computing $C_{\alpha_j}(D) = G(\alpha_j)$ depends, essentially and naturally, on the shape of the threshold sets $C_\alpha(M_i)$ given by the μ_{M_i} according to (2.5). As a special case we consider hypercuboids

(4.1) $$C_{\alpha_j}(M_i) := \{(\underline{x}, y) \in Z : \underline{x}_{ui}(\alpha_j) \leq \underline{x} \leq \underline{x}_{0i}(\alpha_j); \; y_{ui}(\alpha_j) \leq y \leq y_{0i}(\alpha_j)\} .$$

In this case we can use, because of the continuity of f on B, a result due to SCHMERLING (1980):

(4.2) $$G_i(\alpha_j) = \{\underline{\vartheta} \in \Theta : y_{ui}(\alpha_j) \leq \max_{\underline{x} \in Q_{xi}(\alpha_j)} f(\underline{x}, \underline{\vartheta}); \; y_{0i}(\alpha_j) \geq \min_{\underline{x} \in Q_{xi}(\alpha_j)} f(\underline{x}, \underline{\vartheta})\}$$

where $Q_{xi}(\alpha_j) = \{\underline{x} \in B : (\underline{x}, y) \in C_{\alpha_j}(M_i)\}$.

This means that we can obtain the parameter set $G(\alpha_j)$ by solving a certain system of inequalities, in general non-linear with respect to $\underline{\vartheta}$.

If we find the $C_{\alpha_j}(M_i)$ not to be hypercuboids it seems expedient, from the numerical aspect, to try an approximation of $C_{\alpha_j}(M_i)$ by hypercuboids.

Because of the monotonicity of the threshold sets with respect to α and to set inclusion we can stop computation, when for some $\tau \leq T$ we obtain $G(\alpha_\tau) = \emptyset$.

5. Example

The liberation of drug substance y (quantity dissolved in aqua destillata) observed after 8 hours, measured in mg, and the border angle $x_1 = x$, measured in degrees, at the solid/liquid interphase are connected by the relationship

$$(5.1) \qquad y = \vartheta_0 + \vartheta_1 \cos x$$

with $y > 0$ and $0 < x < 180°$, and Θ is defined by

$$(5.2) \qquad 0 < \vartheta_{ou} \leqq \vartheta_o \leqq \vartheta_{oo}$$

$$0 < \vartheta_{1u} \leqq \vartheta_1 \leqq \vartheta_{1o}; \qquad \vartheta_{ou} = \vartheta_{1u}$$

(viz. example in BANDEMER/SCHMERLING, 1983).

From repeated measurement (viz. SCHADLICH, 1973) in five different samples consisting of some basic substance and a drug substance, the portion of which differed from one sample to the other, we specified initial threshold sets $C_0(M_i)$; $i = 1, 2, \ldots, 5$; as rectangles according to (4.1), (cp. table 1). The membership functions μ_i were assumed to be the surfaces of pyramids of height 1 having $C_0(M_i)$ as their bases (s. figure 1). We used $C_{\alpha_j}(M_i)$ with $\alpha_j = 0$; 0.2; 0.4; 0.6; 0.8; 0.898 to calculate $G(\alpha_j) = C_{\alpha_j}(D)$, (cp. figure 2).

Table 1:
corner points of the threshold sets Q_j (0)

j	x_{uj}	x_{oj}	y_{uj}	y_{oj}
1	87.5	95.7	108.8	120.4
2	71.8	83.8	110.8	128.0
3	54.5	65.5	114.2	134.0
4	38.1	49.9	120.5	139.7
5	32.2	44.0	124.9	139.9

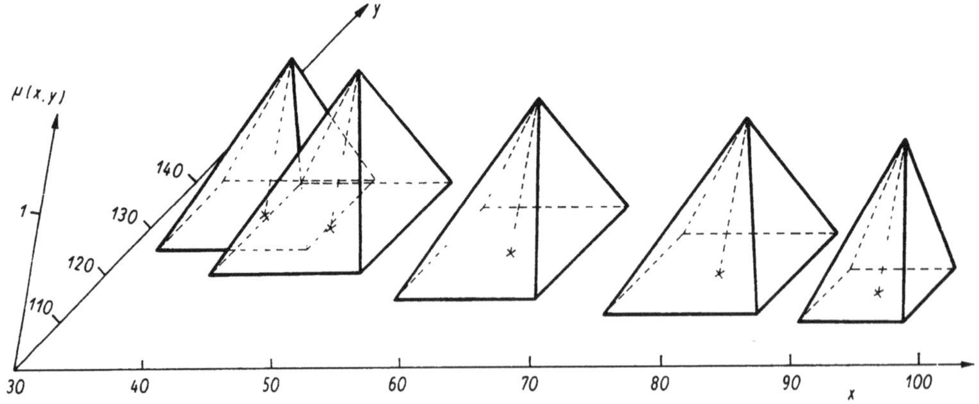

Fig. 1. The membership functions of the fuzzy observations: pyramids with equal height but different area of their bases. Each pyramid represents the fuzzy weakness of the measurements in a charge with given drug substance portion. x is the border angle, measured in degrees, and y is the liberation of drug substance, observed after eight hours, measured in mg.

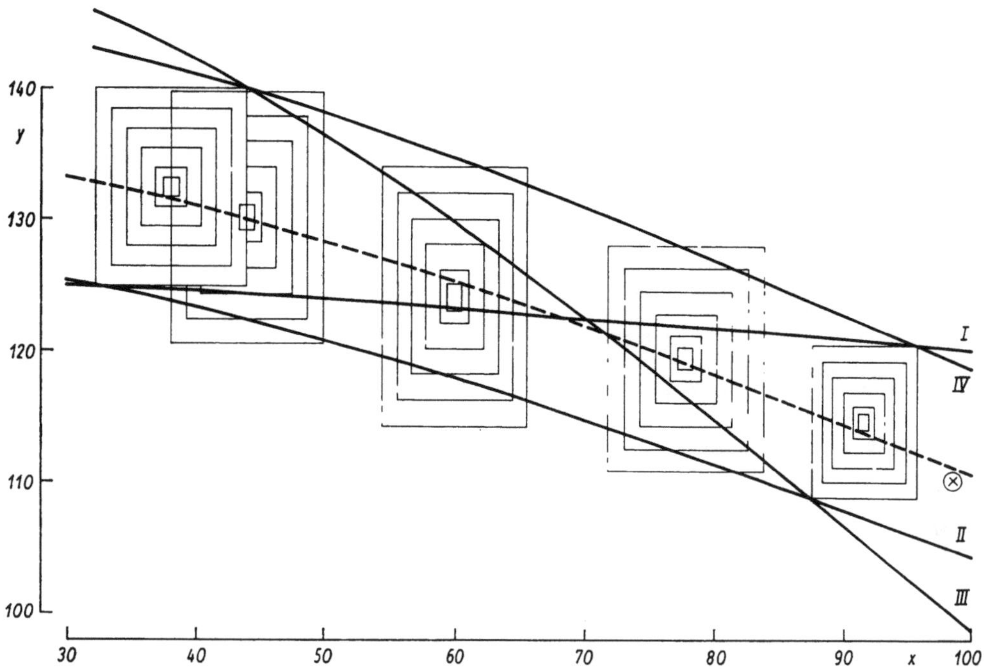

Fig. 2. The threshold sets of the membership functions from Fig. 1 for $\alpha_i=0(0.2)0.8$ and $\alpha_6=0.898$ of each charge (the largest rectangle belongs to α_1 and the smallest one to α_6). The curves I to IV are the boundaries of $G(0)$ according to the corner points I to IV of $G(0)$ in Fig. 3. The curve ⊗ is the boundary of $G(0.898)$, it is the only one for α_6 and corresponds to the point marked by ⊗ in Fig. 3. All the curves are of the form $y=\vartheta_0+\vartheta_1 \cos x$.

According to (4.2) the $G_i(\alpha_j)$ are the solution sets of the system of the following two inequalities

$$(5.3) \qquad y_{ui}+\frac{\alpha_j}{2}(y_{oi}-y_{ui}) \leqq \vartheta_0+\vartheta_1 \cos\left[x_{ui}+\frac{\alpha_j}{2}(x_{oi}-x_{ui})\right]$$

$$y_{oi}-\frac{\alpha_j}{2}(y_{oi}-y_{ui}) \geqq \vartheta_0+\vartheta_1 \cos\left[x_{oi}-\frac{\alpha_j}{2}(x_{oi}-x_{ui})\right]$$

together with the restrictions (5.2).

The parameter set $G(\alpha_j)$, being the intersection of the $G_i(\alpha_j)$ obtained from (5.3) and (5.2), turns out to be a convex polygon (viz SCHMERLING, 1980), which contracts to one point for α_6 (cp. figure 3). The coordinates of the corner points of all $G(\alpha_j)$ are listed in table 2. From figure 3 we can read the value of μ_D for every interesting value $(\vartheta_0, \vartheta_1)$.

In this way we obtain an impression of the grade of validity of our explicit functional relationship for different values of the parameter vector with respect to the totality of our fuzzy observations.

Table 2

corner points of the threshold sets $G(\alpha_i)$

i	$(\vartheta_1, \vartheta_0)$
1	(4.76; 120.87); (20.06; 107.92)
	(46.02; 106.79); (23.82; 122.77)
2	(7.78; 119.90); (20.40; 109.36)
	(40.39; 108.78); (23.41; 121.23)
3	(10.98; 118.86); (20.75; 110.81)
	(35.20; 110.59); (23.00; 119.71)
4	(14.37; 117.73); (21.11; 112.27)
	(30.41; 112.26); (24.93; 116.43)
	(21.30; 118.12)
5	(17.99; 116.52); (21.47; 113.73)
	(25.44; 113.79); (19.64; 116.59)
6	(21.65; 114.46)

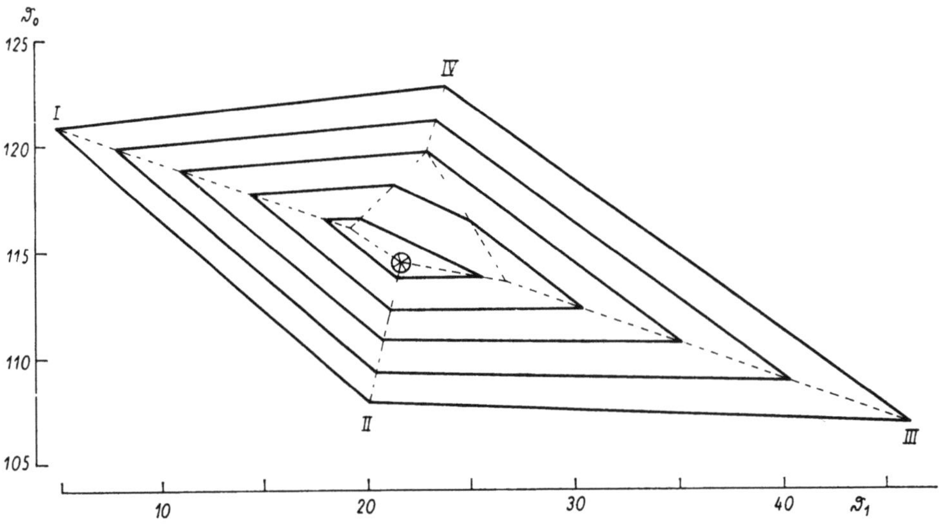

Fig. 3. The level lines of the membership function of the fuzzy parameter set induced by the given fuzzy observations. The lines are drawn with increasing membership values 0(0.2)0.8 beginning with the largest line. The maximum of the membership function is marked by ⊗ and has the value 0.898. The dotted line marks the ridge of the surface projected on ground-plan.

Zusammenfassung

Als eine Kombination von Ideen aus der Intervallarithmetik und der Fuzzytheorie wird angenommen, daß die Beobachtungen, mit denen die explizite Funktionalbeziehung untersucht werden soll, unscharfe Mengen sind, die durch ihre Schwellwertmengen gegeben sind, deren Form die Anwendung intervallarithmetischer Methoden gestattet. Als Ergebnis erhält man eine unscharfe Menge über dem Parameterbereich, aus der man Hinweise auf die Gültigkeit der Funktionalbeziehung erhalten kann. Das Vorgehen wird mit einem Beispiel illustriert.

Резюме

Комбинируя идеи из арифмемпки интервалов и из теории нечёткиx множеств мы предположим, что наблюдения, с помощью которых исследуется явное функциональное соотношение, являются нечеткими множествами. Эти множества даются пороговыми множествами, вид которых делает возможным применение методов из арифметики интервалов. В результате получается нечёткое множество в областе параметров, из которого можно получить указания на действительность функцио-нального соотношения. Метод иллюстрируется примером

Appendix

Lemma A: Under the assumptions of Lemma 2.1 it holds for all $\alpha \in (0, 1]$

$$\mu_{s_n}(f; M_1, \ldots, M_n) \geqq \alpha \Leftrightarrow \forall \quad i \in \{1, \ldots, n\} : \{(\underline{x}, y) \in Z : y = f(\underline{x})\} \cap C_\alpha(M_i) \neq \emptyset$$

Proof: a) $\Rightarrow$: From (2.8) it follows for all $i \in \{1, \ldots, n\}$ that $\sup_{\underline{x} \in B} \mu_{M_i}(\underline{x}, f(\underline{x})) \geqq \alpha$. Because of the assumptions of Lemma 2.1 the supremum is attained by a maximum, i.e. there exists (for each $i \in \in \{1, \ldots, n\}$) some $(\underline{x}^*, f(\underline{x}^*)) \in Z$ being an element of C_α (M_i) according to (2.2) and Definition 2.1.

b) $\Leftarrow$: For each $i \in 1, \ldots, n\}$ there exists some point $(\underline{x}^*, f(\underline{x}^*))$ with $\mu_{M_i}(\underline{x}^*, f(\underline{x}^*)) \geqq \alpha$ and 2.8 completes the proof.

References

Bandemer, H., 1983: Evaluating explicit functional relationships from fuzzy observations. Fuzzy Sets and Systems **15** (1985) 2, 123–130.

Bandemer, H./Schmerling, S., 1983: Construction of estimation regions for the parameters of explicit functional relationships, submitted to Biom. J.

Dubois, D./Prade, H., 1980: Fuzzy sets and systems: Theory and application. Academic Press.

Schädlich, Th., 1973: Zur Bestimmung des Grenzwinkels und seiner Beziehung zur Arzneistofffreigabe, Diplomarbeit, Martin-Luther-Universität Halle-Wittenberg, Sektion Pharmazie (unpublished).

Schmerling, S., 1980: Über die Schätzung von Parametern in expliziten funktionalen Beziehungen auf der Grundlage von Intervalldaten. Dissertation A, Martin-Luther-Universität Halle—Wittenberg, Fakultät für Naturwissenschaften. (unpublished)

Manuscript received: March 24, 1983

Authors' addresses:

Prof. Dr. H. Bandemer
Bergakademie Freiberg
Sektion Mathematik
DDR - 9200 Freiberg
B.-v.-Cotta-Str. 2

Dr. S. Schmerling
Martin-Luther-Universität Halle – Wittenberg
Sektion Mathematik
DDR - 4020 Halle (Saale)
Universitätsplatz 6

Biom. J. 27 (1985) 2, 158

Book Reviews

JEFFERS, J. N. R.: Modelling. Outline Studies in Ecology. Chapman & Hall, London – New York 1982. 80 S., £ 2.75.

Das Büchlein erscheint in der Reihe „Outline Studies in Ecology", die nach Aussage des Herausgebers in kurzer Form mit Hilfe kleiner Bändchen sich stets einer relativ begrenzten Thematik zuwendet, auf diesem Gebiet aber den wissenschaftlich-aktuellen Stand widerspiegeln soll. Daß das mit dem Bändchen „Modelling" gelungen ist, kann man dem Autor bescheinigen. Allein die Kapitel über die zur Verfügung stehenden Mathematischen Modelle, die mitunter kritiklos oder auch blind verwendet wurden sowie die im Anhang angegebene „Modelling checklist" sind von großem Wert. Sie helfen dem „euphorischen" Modellierer, auf dem Boden der Realität zu bleiben und nicht die Welt nach seinem Modell einzurichten. Die einzelnen Kapitel lauten:
1. What is a model? 2. Why do I need a model! 3. How do I start? 4. What help can I expect from mathematics? 5. Do I need a Computer? 6. How do I know when to stop? Appendix: Modelling checklist.
So suggestiv wie die Fragen gestellt sind, so einfach und anschaulich werden die Antworten gegeben, also der Stoff abgehandelt, der einen außerordentlich breiten Leserkreis interessieren dürfte und auch zu empfehlen ist.

H. Ahrens

Biom. J. 27 (1985) 2, 159–166

Use of the Score Statistic for Testing Goodness-of-Fit of Some Generalised Distributions

J. C. W. RAYNER and R. I. MCINTYRE

Mathematics Department, University of Otago, New Zealand

Abstract

The use of the score statistic to test whether a generalised distribution gives an improved fit over a non-generalised distribution is recommended. The score statistic for a generalised exponential family is derived. Several specific examples are given.

Key words: Score statistic, goodness-of-fit, exponential family.

1. Introduction

According to SHOUKRI (1982),
"During the last three decades the statistical literature was enriched by many of the generalised families and classes of discrete distributions, which proved to have many important applications in a wide variety of disciplines, such as biological and medical sciences, social and physical sciences, engineering, operations research, and so on".

Typically a generalised distribution involves more parameters than the simpler, non-generalised distributions from which they are derived. It is not surprising then, that the generalised distribution gives an improved fit. In many articles this improved fit is seen in a reduction of the magnitude of the X^2 goodness-of-fit statistic.

However in most applications a *parsimonious* model is required: no more parameters should be included in the model than are required to adequately describe the data. While X^2 may answer the question of the adequacy of the fit, the question, "does adding an extra parameter to the model significantly improve the fit?" is different. To answer this question, a test based on the score statistic is recommended here.

COX and HINKLEY (1974, Chapter 9) derive the test based on the score statistic and review its development (p. 355). It is perhaps due to this treatment that this test is enjoying a deserved upsurge in usage. See for example KOPECKY and PIERCE (1979), THOMAS and PIERCE (1979), BROWN (1982), and ATKINSON and LAM (1982).

Incidentally the $C(\alpha)$ test of NEYMAN (1959, 1979) is equivalent to that based on the score statistic. FAZAL (1976, 1977) calculated the $C(\alpha)$ tests for two of the distributions of interest here, but generally the calculation of the $C(\alpha)$ tests is less convenient than the following approach.

Finally the focus here is on an exponential family given by JAIN and KHAN (1979). This family includes the general linear exponential distributions (equation (12) of JOHNSON and KOTZ (1982)), which in turn includes the power series and Lagrange distributions (JOHNSON and KOTZ (1982, sections 8 and 11)). Thus the family is both useful and broad.

2. The Score Statistic

For a fuller discussion of the test based on the score statistic and its properties, see COX and HINKLEY (1974), THOMAS and PIERCE (1979) and KOPECKY and PIERCE (1979). From the latter note
"that the quadratic score statistic . . . does note provide a strongly optimal test . . . MORAN (1970), however, demonstrated its asymptotic equivalence to Wald's statistic and the likelihood ratio under regularity conditions. In addition, it is the optimal $C(\alpha)$ test of NEYMAN (1959) and BHAT and NAGNUR (1965) and is widely used."

A random sample $X_1, X_2, \ldots, X_n$ is taken from a population with probability density function $f(x; \underset{\sim}{\theta})$, where $\underset{\sim}{\theta}=(\theta_j)$ is a $k\times 1$ vector from a k dimensional parameter space Ω. Henceforth the regularity conditions of COX and HINKLEY (1974, p. 281) are assumed; this will ensure the existence of the quantities required. Write $l=l(\underset{\sim}{\theta}; x)=\ln\left\{\prod_{j=1}^{n} f(x_j; \theta)\right\}$. The efficient score is then $\underset{\sim}{U}=U(\underset{\sim}{\theta})=(\partial l/\partial\theta_j)$ and the information is $\underline{I}=\underline{I}(\theta)=\mathrm{E}[\underset{\sim}{U}\underset{\sim}{U}^T]=-\mathrm{E}(\partial^2 l/\partial\theta_i\partial\theta_j)$.

Suppose now that $\underset{\sim}{\theta}$ is partitioned into $p\times 1$ and $(k-p)\times 1$ vectors $\underset{\sim}{\psi}$ and $\underset{\sim}{\lambda}$ respectively so that $\underset{\sim}{\theta}^T=(\psi^T, \lambda^T)$. The aim is to test $H: \underset{\sim}{\psi}=\underset{\sim}{\psi}_0$ against $K: \underset{\sim}{\psi}\neq\underset{\sim}{\psi}_0$ with $\underset{\sim}{\lambda}$ as an unspecified nuisance parameter. To do this partition $\underset{\sim}{U}$, $\underline{I}$ and $\underline{I}^{-1}$:

$$\underset{\sim}{U}=\begin{pmatrix}\underset{\sim}{U}_\psi\\ \underset{\sim}{U}_\lambda\end{pmatrix}=\begin{pmatrix}\partial l/\partial\psi\\ \partial l/\partial\lambda\end{pmatrix},\quad \underline{I}=\begin{pmatrix}\underline{I}_{\psi\psi}\underline{I}_{\psi\lambda}\\ \underline{I}_{\lambda\psi}\underline{I}_{\lambda\lambda}\end{pmatrix},\quad \underline{I}^{-1}=\begin{pmatrix}\underline{I}^{\psi\psi}\underline{I}^{\psi\lambda}\\ \underline{I}^{\lambda\psi}\underline{I}^{\lambda\lambda}\end{pmatrix},$$

where, for example, $I_{\psi\psi}=I_{\psi\psi}(\underset{\sim}{\theta})=\mathrm{E}[\underset{\sim}{U}_\psi\underset{\sim}{U}_\psi^T]$. Write $\hat{\underset{\sim}{\psi}}$, $\hat{\underset{\sim}{\lambda}}$, $\hat{\underset{\sim}{\lambda}}_0$, for the maximum likelihood estimates (MLEs) of $\underset{\sim}{\psi}$, $\underset{\sim}{\lambda}$, and $\underset{\sim}{\lambda}$ when $\underset{\sim}{\psi}$ is constrained so that $\underset{\sim}{\psi}=\underset{\sim}{\psi}_0$. Note that $\underline{I}^{\psi\psi}$ is the covariance matrix of the limiting marginal normal distribution of $\hat{\underset{\sim}{\psi}}$, and that $[\underline{I}^{\psi\psi}]^{-1}=\underline{I}_{\psi\psi}-\underline{I}_{\psi\lambda}\underline{I}_{\lambda\lambda}^{-1}\underline{I}_{\lambda\psi}$ (see ANDERSON (1958, p. 42, problem 18)).

The score statistic is $W_u=\underset{\sim}{U}_\psi^T\underline{I}^{\psi\psi}\underset{\sim}{U}_\psi$ where the supressed parameter arguments are $\underset{\sim}{\psi}=\underset{\sim}{\psi}_0$, $\underset{\sim}{\lambda}=\hat{\underset{\sim}{\lambda}}_0$. If a consistent estimator other than $\hat{\lambda}_0$ is used, then $\underset{\sim}{U}_\psi$ must be replaced by $\underset{\sim}{U}_\psi-\underline{I}_{\psi\lambda}\underline{I}_{\lambda\lambda}^{-1}\underset{\sim}{U}_\lambda$, with the parameter arguments being $\underset{\sim}{\psi}=\underset{\sim}{\psi}_0$ and $\underset{\sim}{\lambda}=\underset{\sim}{\lambda}_0^*$, the consistent estimator of $\underset{\sim}{\lambda}$ with $\underset{\sim}{\psi}=\underset{\sim}{\psi}_0$.

If $f(x; (\underset{\sim}{\psi}_0^T, \underset{\sim}{\lambda}^T)^T)$ is a location scale family, then $\underline{I}^{\psi\psi}$ will be $\underset{\sim}{\lambda}$ free. Under H_0 the score statistic has approximately a central χ^2 distribution, χ_p^2. When the null hypothesis is false the distribution is approximately noncentral $\chi^2_{p,\varphi}$, where the parameter of non-centrality is $\varphi = (\underset{\sim}{\psi} - \underset{\sim}{\psi}_0)^T (\underline{I}^{\psi\psi})^{-1} (\underset{\sim}{\psi} - \underset{\sim}{\psi}_0)$. These distributions are derived assuming a sequence of local alternatives; namely $\underset{\sim}{\psi} = \underset{\sim}{\psi}_0 + \underset{\sim}{\delta}_\psi / \sqrt{n}$.

3. An Exponential Family

As in JAIN and KAHN (1979) consider the exponential family of probability density functions $\Phi^x a(x; u) f^{-u}$ where $f = f(\varrho)$ and $\Phi = \Phi(\varrho)$ are independent of u. Let mu be the mean of this family. Another exponential family may be generated from this family by

$$(3.1) \qquad g(x; r, u, \varrho) = \frac{u}{u + rx} \Phi^x a(x; u + rx) f^{-(u+rx)},$$

$$\text{where} \quad x > 0, \quad r \geqq 0, \quad mu\, r < 1 .$$

Putting $r = 0$ recovers the original distribution.

Whether or not $g(x; 0, u, \varrho)$ or $g(x; r, u, \varrho)$ adequately fits data of interest, it may be necessary to determine if $g(x; r, u, \varrho)$ significantly improves the fit provided by $g(x; 0, u, \varrho)$. Occasionally it will be useful to answer this question conditional on either $u = u_0$ or $\varrho = \varrho_0$. Initially then we test $H: r = 0$ against $\mathrm{K}: r \neq 0$ with both u and ϱ nuisance parameters. In the notation of the previous section $\psi \equiv r$, $\lambda^T \equiv (u, \varrho)$.

Write $a' = a'(x; u) = \partial\{a(x; y)\}/\partial y \mid y = u$, $d\Phi/d\varrho = \Phi_\varrho$ and $df/d\varrho = f_\varrho$. Then the conditional MLEs ($\hat{u}_0$, $\hat{\varrho}_0$, equivalent to $\hat{\lambda}_0$) are the solutions to

$$\begin{cases} \left\{\sum_{i=1}^n a'(x_i; u)/a(x_i; u)\right\} - n \ln f = 0 \\ \bar{x}\Phi_\varrho/\Phi - u f_\varrho/f = 0 . \end{cases}$$

The efficient score is

$$\underset{\sim}{U} = \begin{pmatrix} \left\{\sum_{i=1}^n x_i a'/a\right\} - n\bar{x}\,[(1/\hat{u}_0) + \ln f] \\ \left\{\sum_{i=1}^n a'/a\right\} - n \ln f \\ n[\bar{x}\Phi_x/\Phi - \hat{u}_0 f_\varrho/f] \end{pmatrix}$$

where the supressed arguments are $r = 0$, $u = \hat{u}_0$, $\varrho = \hat{\varrho}_0$. To write the information matrix compactly put $f^* = (f_\varrho/f)^2 - (d^2f/d\varrho^2)/f$, $\Phi^* = (\Phi_\varrho/\Phi)^2 - (d^2\Phi/d\varrho^2)/\Phi$ and $a^* = (a'/a)^2 - a''/a$ where $a'' = \partial^2\{a(x; y)\}/\partial y \mid y = u$. Of course the supressed arguments are as for $\underset{\sim}{U}$.

$$(3.2) \quad n^{-1}\underline{I} = n^{-1}\underline{I}(0, \hat{u}_0, \hat{\varrho}_0) = \begin{pmatrix} \mathrm{E}\,[X^2\,(a^* - 1/u^2)] & \mathrm{E}\,[X\,(a^* - 1/u^2)] & \mathrm{E}\,[X f_\varrho/f] \\ \mathrm{E}\,[X(a^* - 1/u^2)] & \mathrm{E}[a^*] & f_\varrho/f \\ \mathrm{E}\,[X f_\varrho/f] & f_\varrho/f & u f^* - \Phi^*\mathrm{E}[X] \end{pmatrix}.$$

By noting that $E[\underset{\sim}{U}]=\underset{\sim}{O}$ and differentiating this equation it is possible to obtain alternative forms for some elements of the above matrix. Some helpful identities are

$$E[a'/a]=\ln f; \qquad E[X]=uf_\varrho\Phi/\{f\Phi_\varrho\}=\mu \text{ say :}$$
$$E[Xa'/a]=\mu\{1/u+\ln f\};$$
$$E[Xa^*]=E[X(a'/a)^2]-\mu(1+u\ln f)(\ln f)/u.$$

In these expectations $r=0$. In the examples that follow the form (3.2) of $\underline{I}$ has been preferred to other alternatives. However for examples such as 4 (iv) the above identities may be helpful.

4. Examples

(i) The quasi-binomial distribution

This distribution has probability function

$$g(x;\psi,\lambda)=\binom{u}{x}\lambda(1-\lambda)(\lambda+x\psi)^{x-1}(1-\lambda+u\psi-x\psi)^{u-x-1}(1+u\psi)^{1-u},$$

$$(4.1) \qquad x=0,1,\ldots,u; \quad 0\leqq\lambda\leqq 1, \quad \text{and} \quad 0\leqq\psi<1.$$

It was first introduced by CONSUL (1974) and then further discussed and refined by CONSUL and MITTAL (1975) and CONSUL (1975). FAZAL (1976) calculated the $C(\alpha)$ test. For a sample of size n, the test statistic is

$$Z_n=(\bar{X}^2-u\bar{X}+uS^2)/[\bar{X}(u-\bar{X})]\sqrt{[nu/2(u-1)]}$$

$$\text{where} \quad nS^2=\sum_{i=1}^{n}(X_i-\bar{X})^2, \qquad \bar{X}=\sum_{i=1}^{n}X_i/n$$

and asymptotically Z_n has the $N(0,1)$ distribution.

Although (4.1) is not of the form (3.1), the calculation of the score statistic is straightforward. MCINTYRE (1982) showed that $W_u=Z_n^2$, as expected. In addition he simulated the null distribution of W_u for various combinations of $u=5$, 10, 20, $\varrho=.05$, .1, .2, .5, and $n=10$, 20, 30. He concluded that
"these results seem to indicate a bumpiness for the distributions for small values of u, ϱ and n and a gradual smoothing as u and n increase and ϱ tends to 0.5. Thus it would appear that W_u does follow the χ_1^2 distribution, specifically when u and n are large and ϱ is close to 0.5." (MCINTYRE (1982, p. 35))

(ii) The generalized Poisson distribution

CONSUL and JAIN (1970, 1973) introduced the generalized Poisson distribution, the so called Borel-Tanner distribution (JAIN (1974)), with probability function,

$$g(x;\psi,\lambda)=\lambda(\lambda+x\psi)^{x-1}e^{-(\lambda+x\psi)}/x!,$$
$$x=0,1,2,\ldots,\lambda>0, \qquad |\psi|<1, \quad \text{and with}$$

$$(4.2) \qquad g(x;\psi,\lambda)=0 \quad \text{for} \quad x\geqq m \quad \text{if} \quad \lambda+m\psi\leqq 0.$$

JAIN and CONSUL (1971) derived (4.2) as the limiting form of a generalized negative binomial distribution. FAZAL (1977) derived the $C(\alpha)$ test.

In (3.1) put $\Phi(\varrho)=\varrho$, $f(\varrho)=\exp(\varrho)$, $a(x;u)=u^x/x!$. The resulting family apparently has three parameters, but in reality has only two: ϱu and ϱr. Putting $\varrho=1$ gives a form equivalent to (4.2). In the derivations in section three omit all rows and columns relating to differentiation by ϱ. Then $\hat{u}_0=\bar{X}$,

$$\underset{\sim}{U}^T=(\{\sum X_i^2/\bar{X}\}-n-n\hat{N},\,0)=(n\,(S^2-\bar{X})/\bar{X},0) \quad \text{if} \quad nS^2=\sum X_i^2-n\bar{X}^2,$$

and

$$\underline{I}(0,u)=\frac{n}{u^2}\begin{pmatrix}\mathrm{E}\,[X^2\,(X-1)] & \mathrm{E}\,[X\,(X-1)]\\ \mathrm{E}\,[X\,(X-1)] & \mathrm{E}\,[X]\end{pmatrix}=\frac{n}{u}\begin{pmatrix}u^2+2u & u\\ u & 1\end{pmatrix}.$$

Thus

$$\underline{I}^{-1}(0,u)=\begin{pmatrix}1 & u\\ u & u^2+2u\end{pmatrix}/(2n).$$

Since

$$\underline{I}^{rr}=1/2n, \qquad W_u=\{(S^2/X)-1\}^2\,n/2\,.$$

FAZAL (1977) gives several examples in most of which the extra parameter of the generalized Poisson significantly improves the fit given by the ordinary Poisson. MCINTYRE (1982) examined the null distribution of W_u. He found χ_1^2 to be a good approximation, with the distribution being "bumpy" for small u and n, "smoothing out" as u and n increase.

(iii) A generalized gamma distribution

KHAN and JAIN (1978) and JAIN and KHAN (1979) discuss the generalized gamma distribution with probability function

(4.3) $$g(x;r,u,\varrho)=u\exp(-x/\varrho)\,x^{u+rx-1}/\{\varrho^{u+rx}\Gamma(u+rx+1)\}, \quad x>0\,,$$ provided $r\varrho<1$.

This is of the form (3.1) with $\Phi(\varrho)=\exp(-1/\varrho)$, $f(\varrho)=\varrho$, and $a(x;u)=x^{u-1}/\Gamma(u)$. With a three parameter family there are a variety of hypotheses that may be of interest. KHAN and JAIN (1978) consider an example with $u=1$ where the extra parameter r reduces X^2 from 21.1576 to 9.5578. It may be of interest to test $(r,u)=(0,1)$ with ϱ as a nuisance parameter or to test say $r=0$ with both u and ϱ as nuisance parameters. Here only the latter is considered.

Given Φ, f, a as above we quickly find that $a'/a=\ln x-\varphi(u)$, where φ is the psi or digamma function (see ABRAMOWITZ and STEGUN (1970)), $a^*=\varphi'(u)$, the trigamma function, $\Phi_\varrho/\Phi=\varrho^{-2}$, $f_\varrho/f=\varrho^{-1}$. The constrained maximum likelihood estimates $\hat{u}_0$, $\hat{\varrho}_0$ are the solutions to

$$\begin{cases}\sum \ln x_i-n\varphi(u)-n\ln\varrho=0\\ n\bar{x}/\varrho^2-nu/\varrho=0\,.\end{cases}$$

U_r, the component of the efficient score corresponding to differentiation by r, is given by

$$U_r = \sum x_i \ln x_i \quad n\bar{x}\,(1/u + \varphi(u) + \ln \varrho)$$
$$= \sum x_i \ln x_i - n\bar{x}\,(\varphi\,(u+1) + \ln \varrho)$$ by (6.3.5) of ABRAMOWITZ and STEGUN (1970).

The information matrix satisfies

$$I(0, u, \varrho)/n = \begin{pmatrix} (\varphi'(u) - u^{-2})\,\mathrm{E}[X^2] & (\varphi'(u) - u^{-2})\,\mathrm{E}[X] & \mathrm{E}[X]/\varrho \\ (\varphi'(u) - u^{-2})\,\mathrm{E}[X] & \varphi'(u) & 1/\varrho \\ \mathrm{E}[X]/\varrho & 1/\varrho & 2\mathrm{E}[X]/\varrho^3 - u\varrho^2 \end{pmatrix}.$$

By (6.4.6) of ABRAMOWITZ and STEGUN $\varphi'\,(z+1) = \varphi'(z) - z^{-2}$. Also when $r = 0$, $\mathrm{E}\,[X] = u\varrho$, $\mathrm{E}\,[X^2] = u\,(u+1)\,\varrho^2$, so that

$$I(0, u, \varrho)/n = \begin{pmatrix} u\,(u+1)\,\varrho^2\varphi'\,(u+1) & u\varrho\varphi'\,(u+1) & u \\ u\varrho\varphi'\,(u+1) & \varphi'(u) & 1/\varrho \\ u & 1/\varrho & u/\varrho^2 \end{pmatrix}.$$

This matrix has determinant $u^2\{\varphi'(u)\}^2 - \varphi'(u) - 1 = \det$ say, so that its (1,1)th element is $\{u\varphi'(u) - 1\}/[\varrho^2 \det]$. It follows that

$$W_u = \frac{\{\sum X_i \ln X_i - n\bar{X}(1 - \gamma + \ln \hat{\varrho}_0)\}^2\,\{\hat{u}_0\varphi'(\hat{u}_0) - 1\}}{n\hat{\varrho}_0^2\,[\hat{u}_0^2\{\varphi'(\hat{u}_0)\}^2 - \varphi'(\hat{u}_0) - 1]}, \tag{4.4}$$

where γ is Euler's constant, defined by $\gamma = \lim_{m\to\infty} \left[\sum_{r=1}^{m} \frac{1}{r} - \ln m\right]$, with $\gamma \approx .57721566$.

If the model specifies $u = 1$ a similar development gives

$$W_u = \{\sum X_i \ln X_i - n\bar{X}\,(1 - \gamma + \ln \bar{X})\}^2\, 3/\{n\,(\pi^2 - 9)\,\bar{X}^2\}. \tag{4.5}$$

For the survival data of Table 2 of KHAN and JAIN (1978) the statistic (4.5) has the value 5.631465, indicating at the 5 % level (but not at the 1 % level) that the generalized exponential model improves the fit given by the ordinary exponential.

(iv) A generalized negative binomial distribution

This distribution was discussed by JAIN and CONSUL (1971), and JAIN (1972, 1975). It has probability function

$$g(x; r, u, \varrho) = u\Gamma\,(u + rx)/\{x!\Gamma\,(u + rx - x + 1)\,\varrho^x\,(1 - \varrho)^{u + rx - x},$$
$$x = 0, 1, 2, \ldots \text{ where } g(x; r, u, \varrho) = 0 \quad \text{for} \quad x \leqq m \quad \text{if}$$
$$u + rx < 0, \quad \text{and} \quad 0 < \varrho < 1, \quad |r\varrho| < 1\,. \tag{4.6}$$

This is of the form (3.1) with $\Phi(\varrho) = \varrho/(1 - \varrho)$, $f(\varrho) = 1/(1 - \varrho)$, and $a(x; u) = \binom{u}{x}$. To test $H: r = 0$ (that is, the ordinary binomial distribution) we need

$$a'/a = \varphi\,(u+1) - \varphi\,(u - x + 1), \qquad a^* = \varphi'\,(u - x + 1) - \varphi'\,(u+1)\,,$$
$$\Phi_\varrho/\Phi = 1/[\varrho\,(1 - \varrho)], \qquad f_\varrho/f = 1/(1 - \varrho)\,.$$

$\hat{u}_0$, $\hat{\varrho}_0$ are the solutions to

$$\begin{cases} \varphi(u+1) - \sum_{i=1}^{n} \varphi(u - x_i + 1)/n + \ln(1-\varrho) = 0 \\ u\varrho - \bar{x} = 0\,. \end{cases}$$

Further $U_r = n\bar{X}\,\{\ln(1-\varrho) - 1/u - \Phi(u+1)\} - \sum X_i \Phi(u - X_i + 1)$, and $I(0, u, \varrho)/n$ is given by

$$\begin{pmatrix} \mathrm{E}\,[X^2\varphi'(u-X+1)] & \mathrm{E}\,[X\varphi'(u-X+1)] & u\varrho/(1-\varrho) \\ -u\varrho\,\{1-\varrho+u\varrho\}\,\{\varphi'(u+1)+u^{-2}\} & -u\varrho\,\{\varphi'(u+1)+u^{-2}\} & \\ \mathrm{E}\,[X\varphi'(u-X+1)] & \mathrm{E}\,[\varphi'(u-X+1) & 1/(1-\varrho) \\ -u\varrho\,\{\varphi'(u+1)+u^{-2}\} & -\{\varphi'(u+1)+u^{-2}\} & \\ u\varrho/(1-\varrho) & 1/(1-\varrho) & u/[\varrho\,(1-\varrho)] \end{pmatrix}.$$

Simplification of the elements of this matrix may be attempted by using the techniques mentioned at the conclusion of the previous section, or by exploiting properties of the particular distribution under consideration. Alternatively $\mathrm{E}\,[f(X)]$ may be approximated by $\sum_{i=1}^{n} f(x_i)/n$; under suitable regularity conditions the latter approaches the former in probability. Having evaluated or approximated $I(0, u, \varrho)/n$, W_u may be obtained as previously. (Its form is too cumbersome to write in full.)

Finally note that if $r=1$ in (4.6) the usual negative binomial probability function is obtained. By reparametrixing (putting $R=r-1$) a distribution of the form (3.1) is obtained. The W_u test for $R=0$, or equivalently $r=1$, may then be obtained as previcusly.

Acknowledgement

The authors wish to acknowledge the contribution of their late colleague and mentor, Gopi Chand Jain, a profilic researcher in the area of distribution theory. The problem addressed in this article, how to test for the fit of the generalised distributions, was posed by Dr. Jain.

Summary

In this article we have argued that the score statistic is appropriate for testing for an improvement in fit that may be provided by using some generalized distributions. An outline is given of the theory of the score statistic, which is derived for an exponential family of JAIN and KHAN (1979). In some cases the statistic is difficult to obtain in a convenient closed form; in others there is no problem, and in some cases we have confirmed the asymptotic null distribution by simulation.

References

ABRAMOWITZ, M., and STEGUN, I. A., 1970: Handbook of mathematical functions. Dover Press, New York.

ANDERSON, T. W., 1958: An introduction to multivariate statistical analysis. Wiley, New York.

ATKINSON, A. C., and LAM YEH., 1982: Inference for Sichel's compound Poisson distribution. J. Amer. Statist. Assoc. 77, 153–158.

BHAT, B. R., and NAGNUR, B. N., 1965: Locally asymptotically most stringent tests and Lagrangian multiplier tests of linear hypotheses. Biometrika 52, 459–468.

Brown, C. C., 1982: On a goodness of fit test for the logistic model based on score statistics. Commun. Statist.-Theor. Meth. **11**, 1087–1105.

Consul, P. C., 1974: A simple urn model dependent upon predetermined strategy. Sankhya, Series B, **36**, 391–399.

Consul, P. C., 1975: On a characterization of Lagrangian, Poisson and quasi-binomial distributions. Commun. Statist. **4**, 555–563.

Consul, P. C., and Jain, G. C., 1970: On the generalisation of Poisson distribution (Abstract). Ann. Math. Statist. **41**, 1386–1387.

Consul, P. C., and Jain, G. C., 1973: A generalisation of the Poisson distribution. Technometrics **15**, 791–799.

Consul, P. C. and Mittal, S. P., 1975: A new urn model with predetermined strategy. Biom. Z. **17**, 67–75.

Cox, D. R. and Hinkley, D. V., 1974: Theoretical statistics, Chapman and Hall, London.

Fazal, S. S., 1976: A test for quasi-binomial distribution. Biom. Z. **18**, 619–622.

Fazal, S. S., 1977: A test for a generalised Poisson distribution. Biom. J. **19**, 245–251.

Jain, G. C., 1972: Approximation of functions by a new class of linear operators. J. Austral. Math. Soc. **13**, 271–276.

Jain, G. C., 1974: A Borel-Tanner distribution and its approximation to the negative binomial distribution. Math. Operationsforsch. Statist. **5**, 207–212.

Jain, G. C., 1975: On power series distributions associated with Lagrange expansion. Biom Z. **17**, 85–97.

Jain, G. C., and Consul, P. C., 1971: A generalised negative binomial distribution. Siam. J. Appl. Math. **21**, 501–513.

Jain, G. C. and Khan, M. S. H., 1979: On an exponential family. Math. Operationsforsch. Statist. **10**, 153–168.

Johnson, N. L., and Kotz, S., 1982: Developments in discrete distributions, 1969–1980. Inter. Statist. Rev. **50**, 71–101.

Khan, M. S. H., and Jain, G. C., 1978: A class of distributions in the first emptiness of a semi-infinite reservoir. Biom. J., **20**, 243–251.

Kopecky, K. J., and Pierce, D. A., 1979: Efficiency of smooth goodness of fit tests. J. Amer. Statist. Assoc. **74**, 393–397.

McIntyre, R. I., 1982: The score statistic. Unpublished Postgraduate Diploma project, Department of Mathematics, University of Otago.

Moran, P. A. P., 1970: On asymptotically optimal tests of composite hypotheses. Biometrika, **57**, 47–55.

Neyman, J., 1959: Optimal asymptotic tests of composite statistical hypotheses. In Probability and Statistics, ed. U. Gnenander, Almqvist and Wiksell, Stockholm.

Neyman, J., 1979: $C(\alpha)$ tests and their use. Sankhya, Series A, **41**, 1–21.

Shoukri, M. M., 1982: On a generalisation for the double Poisson distribution. Commun. Statist.–Theor. Meth. **11**, 151–164.

Thomas, D. R., and Pierce, D. A., 1979: Neyman's smooth goodness-of-fit test when the hypothesis is composit. J. Amer. Statist. Assoc. **74**, 441–445.

Manuscript received: Jan. 13, 1983

Authors' address:

I. C. W. Rayner; R. I. McIntyre
Mathematics Department,
University of Otago,
P. O. Box 56,
Dunedin,
New Zealand.

Biom. J. 27 (1985) 2, 167–172

A Comparison of the MPN and Fisher-Yates Estimators for the Density of Organisms

D. J. BEST and J. C. W. RAYNER

CSIRO Division of Food Research, North Ryde, Australia

Abstract

The method of moments estimator, discussed in FISHER and YATES (1970), for the density of bacteria is compared with the maximum likelihood or MPN estimator. For sample sizes used in practice the estimators are found to be very similar. Estimators which reduce bias are discussed and their use recommended.

Key words: Dilution method, simulation comparisons, bias, confidence limits.

1. Introduction

The dilution method is commonly employed by microbiologists to estimate the density of organisms present in a medium. The method is used when the presence or absence of the organism can be determined and it has a number of important applications concerned with food and health standards.

Suppose k dilutions containing concentrations $z_1, z_2, \ldots, z_k$ of original material are prepared. At the ith dilution n_i tubes or plates are inoculated. The dilution and inoculation processes are performed so that individual responses are independent. The distribution of organisms in the medium is assumed random and hence the number of organisms in a small part of the medium follows a Poisson distribution. If λ specifies the density of organisms in the original medium the probability of a tube or plate inoculated with a dilution of concentration z_i being sterile is thus $e^{-\lambda z_i}$. Suppose y_i is the number of sterile tubes or plates at the ith dilution. Then the likelihood of an observed set of sterile and fertile results is $\prod_{i=1}^{k} \binom{n_i}{y_i} (e^{-\lambda z_i})^{y_i} (1 - e^{-\lambda z_i})^{n_i - y_i}$ and the expected number of sterile tubes is $\sum_{i=1}^{k} n_i e^{-\lambda z_i}$. MCCRADY (1915) suggested λ be estimated using what microbiologists call the MPN estimator (it is, in fact, the maximum likelihood estimator) while FISHER and YATES (1970) suggest estimating λ by solving the equation formed by equating observed and expected numbers of sterile tubes.

The MPN estimator is probably the more commonly used estimator and has desirable asymptotic properties. The Fisher-Yates estimator is, however, still employed (see, for example, BAXTER (1979)) and has the advantage that tables such as Table VIII 2 of FISHER and YATES (1970) apply to any k. Which of the two estimators should be used in practice? As the numbers of tubes or plates used are hardly ever large the Fisher-Yates estimator may compare well with the MPN estimator. In this note non-asymptotic comparisons will be made between Fisher's estimator and the MPN estimator.

2. Comparison of the Estimators

(i) Definition

The Fisher-Yates estimator, $\hat{\lambda}$ say, will be defined to be the solution of the equation

$$\sum_{j=1}^{k} (n_j e^{-\hat{\lambda} z_j} - y_j) = 0 .$$

Note that the estimator actually described in FISHER and YATES (p. 8, 1970) is defined for a more restricted set of n_j, z_j values. The MPN estimator, $\hat{\hat{\lambda}}$ say, will be defined as the solution of the equation

$$\sum_{j=1}^{k} [(n_j - y_j)\, z_j e^{-\hat{\hat{\lambda}} z_j} (1 - e^{-\hat{\hat{\lambda}} z_j})^{-1} - y_j z_j] = 0 .$$

(ii) *First Two Moments*

For given z_i, n_i, k and λ dilution experiments can be simulated by use of a convenient pseudo-random binomial variate generator. The binomial parameters are (n_i, p_i) where $p_i = e^{-\lambda z_i}$. As this study is concerned with practical cases where n_i is small the algorithm BU of AHRENS and DIETER (1974) was employed. For a given realization $y_1, \ldots, y_k$ the IMSL (1980) routine ZXREALI was used to obtain $\hat{\lambda}$ and $\hat{\hat{\lambda}}$ (the routine ZXREALI sometimes needed better starting values when calculating $\hat{\hat{\lambda}}$ than were needed for the calculation of $\hat{\lambda}$). In this way 1,000 estimates for the same (z_i, n_i, k, λ) set were produced. The expectation and variance of $\hat{\lambda}$ and $\hat{\hat{\lambda}}$ were then estimated by the sample mean and variance. The results for $n_1 = n_2 = n_3 = 5$, $z_1 = .1$, $z_2 = .01$, $z = .001$ and various λ are shown in Table 1(a).

Note that the accuracy of the Monte Carlo evaluation of the expected values can be estimated by using the Monte Carlo values for the variances. Thus var $(\mathrm{E}\,(\hat{\lambda})) = (\mathrm{var}\,(\hat{\lambda}))/1\,000$ and so, for example, the variance of the first $\mathrm{E}(\hat{\lambda})$ values in Table 1(a), i.e. 5.68, is approximately 19.19/1 000 or 0.02. Clearly on reference to Table 1 there is little to choose between $\hat{\lambda}$ and $\hat{\hat{\lambda}}$. Other similar simulation studies, not given here, for different n_i, z_i, λ choices were in agreement with this conclusion.

Table 1

(a) Monte Carlo expectations and variances of $\hat{\lambda}$ and $\hat{\hat{\lambda}}$ when $n_1 = n_2 = n_3 = 5$, $z_1 = .1$, $z_2 = .01$ and $z_3 = .001$.

λ	$E(\hat{\lambda})$	var $(\hat{\lambda})$	$E(\hat{\hat{\lambda}})$	var $(\hat{\hat{\lambda}})$
5	5.68	19.99	5.65	14.88
10	11.65	56.00	11.31	52.94
25	29.67	332.74	32.08	389.88
50	59.30	1500.80	60.52	1747.42
75	86.70	3607.69	92.77	3828.04
100	119.27	7157.95	118.59	5601.54

(b) Monte Carlo expectations and variances of $\hat{\lambda}^*$ and $\hat{\hat{\lambda}}^*$ when $n_1 = n_2 = n_3 = 5$, $z_1 = .1$, $z_2 = .01$ and $z_3 = .001$.

λ	$E(\hat{\lambda}^*)$	var $(\hat{\lambda}^*)$	$E(\hat{\hat{\lambda}}^*)$	var $(\hat{\hat{\lambda}}^*)$
5	5.02	14.05	5.01	10.55
10	9.94	36 07	9.66	33.61
25	24.52	249.58	24.53	235.37
50	50.03	1091.48	48.68	1082.01
75	73.15	2330.00	72.35	2006.06
100	99.73	4428.77	95.54	3116.04

The agreement between $\hat{\lambda}$ based on $\sum_{i=1}^{k} y_i$ and $\hat{\hat{\lambda}}$ based on $y_1, y_2, \ldots, y_k$ can be partly explained by the fact that, as expected, the observed outcomes from the simulation experiments were often ordered, i.e. $y_k > y_{k-1} > \ldots > y_1$. For example, when $\lambda = 50$ in Table 1, about 80 % of outcomes are ordered. Given that the y_i are ordered, the value of $\sum_{i=1}^{k} y_i$ is often enough to deduce what the individual y_i are, particularly for small k and n_i. Thus if $k=3$, $n_1 = n_2 = n_3 = 5$ and $y_3 > y_2 > y_1$, $\sum_{i=1}^{3} y_i = 8$ (i.e. $\hat{\lambda} \doteqdot 75$) then the possibilities for the result (y_1, y_2, y_3) are (5, 2, 1), (4, 3, 1) or (5, 3, 0). Likelihood calculations show that (5, 3, 0) and (5, 2, 1), i.e. $\hat{\hat{\lambda}} \doteqdot 79$ and $\hat{\hat{\lambda}} \doteqdot 70$, are much more likely than (4, 3, 1), i.e. $\hat{\hat{\lambda}} \doteqdot 33$.

(iii) *Bias Reduction*

Table 1 also suggests that the bias in $\hat{\lambda}$ or $\hat{\hat{\lambda}}$ may be considerable and SALAMA et al (1978) have proposed corrections, $\hat{\lambda}_1$ and $\hat{\lambda}_2$, to make $\hat{\lambda}$ less biased. Using a Taylor series method it can be shown that if $\tilde{\lambda}$ denotes either $\hat{\lambda}$ or $\hat{\hat{\lambda}}$ a less biased estimate is

$$\tilde{\lambda}^* = \left\{\tilde{\lambda} - 1/2 \sum_{i=1}^{k} \left[\left(\frac{\partial^2 \tilde{\lambda}}{\partial y_i^2}\right) n_i e^{-\tilde{\lambda} z_i} (1 - e^{-\tilde{\lambda} z_i})\right]\right\}.$$

When $\tilde{\lambda}=\hat{\lambda}$,

$$\frac{\partial^2\hat{\lambda}}{\partial y_i^2} = -\left(\sum_{j=1}^{k} n_j z_j^2 e^{-\hat{\lambda} z_j}\right)\left(\frac{\partial\hat{\lambda}}{\partial y_i}\right)^3, \qquad i=1, \dots, k$$

and

$$\frac{\partial\hat{\lambda}}{\partial y_i} = -\left(\sum_{j=1}^{k} n_j z_j e^{-\hat{\lambda} z_j}\right)^{-1}, \qquad i=1, \dots, k\,.$$

Simulation results given in Table 1(b) indicate the improved performance of $\hat{\lambda}^*$ $\hat{\lambda}_1$ $\hat{\lambda}_2=\hat{\hat{\lambda}}^*$. Results not given here indicate that $\hat{\lambda}_2$ was slightly superior to $\hat{\lambda}_1$, as regards bias reduction.

It was observed from the simulations, as may otherwise be obvious, that (a) the sampling distribution of $\hat{\lambda}$ is positively skewed, (b) $\hat{\lambda}^*<\hat{\lambda}$ and (c) $\hat{\lambda}-\hat{\lambda}^*$ increases as $\hat{\lambda}$ increases. Thus $\hat{\lambda}^*$ has a more symmetric distribution than $\hat{\lambda}$ and, as can be seen from Table 1(b), var $(\hat{\lambda}^*)<$ var $(\hat{\lambda})$. The same observations were made for $\hat{\hat{\lambda}}$ and $\hat{\hat{\lambda}}^*$.

(iv) *Confidence Interval Comparisons*

It is also possible to compare $\hat{\lambda}$ and $\hat{\hat{\lambda}}$ by looking at their, say 95 %, confidence intervals. To use published exact 95 % confidence intervals we consider the case $n_1=n_2=n_3=10$ and $z_1=1$, $z_2=.1$, $z_3=.01$. As a given experimental outcome (y_1, y_2, y_3) gives different $\hat{\lambda}$ and $\hat{\hat{\lambda}}$ values the associated intervals will, of course, also differ. However if the upper bounds of intervals and also the lower bounds are joined to form piecewise linear curves and the results graphed as shown in Figure 1(a) then comparisons can be made. Although the intervals for $\hat{\hat{\lambda}}$ are certainly

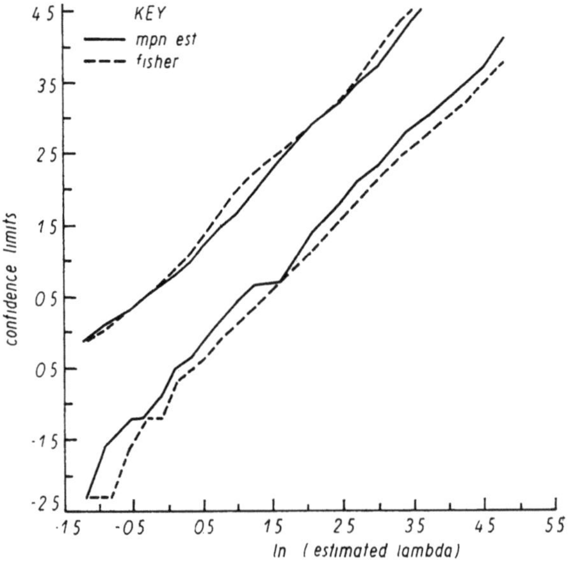

Fig. 1 (a). Exact 95 % confidence intervals

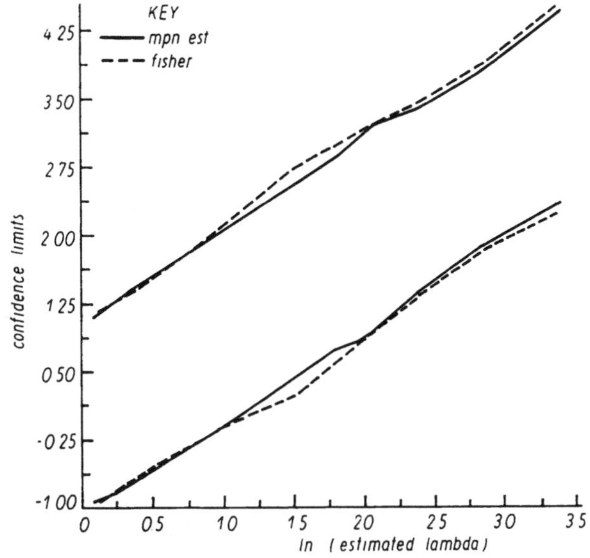

Fig. 1 (b). Approximate 95 % intervals

shorter it is the upper bound which is usually important and for this bound $\hat{\hat{\lambda}}$ is only slightly better than $\hat{\lambda}$. The exact confidence intervals for the parameters stated above are given by DE MAN (1975) for $\hat{\hat{\lambda}}$ and by MATUSZEWSKI et al (1935) for $\hat{\lambda}$.

However, in practice, confidence intervals are often calculated using the approximation that log ($\hat{\hat{\lambda}}$) and log ($\hat{\lambda}$) have a normal distribution (see respectively FISHER and YATES (1970, p. 9) and PARNOW (1972)). PARNOW's (1972) 95 % interval for $\hat{\hat{\lambda}}$ is exp (ln $\hat{\hat{\lambda}}$) $\pm 1.96\, s/\hat{\hat{\lambda}}$) where s is the standard error of $\hat{\hat{\lambda}}$ as given by HALDANE (1939), i.e. $s^2 = \left[\sum_{i=1}^{k} (n_i - y_i)\, z_i^2 e^{z_i\hat{\hat{\lambda}}} \left(e^{z_i\hat{\hat{\lambda}}} - 1\right)^{-2}\right]^{-1}$. Using the usual Taylor series argument $v^2 = \text{var}\,(\hat{\lambda}) = \sum_{i=1}^{k} \left(\frac{\partial\hat{\lambda}}{\partial y_i}\right)^2 n_i e^{-\hat{\lambda} z_i} (1 - e^{-\hat{\lambda} z_i})$. Thus an approximate 95 % interval for $\hat{\lambda}$ is exp (ln ($\hat{\lambda}$) $\pm 1.96\, v/\hat{\lambda}$). These approximate intervals can be shown to agree well with the exact intervals and comparisons based on them again demonstrate that the two sets of confidence limits are not very different. Figure 1(b) shows one such comparison and is for the case $n_1 = n_2 = n_3 = 5$, $z_1 = 1.0$, $z_2 = .1$, $z_3 = .01$.

3. Conclusion

Non-asymptotic comparisons of $\hat{\lambda}$ and $\hat{\hat{\lambda}}$ were made. Monte Carlo estimates of the expected values and variances of $\hat{\lambda}$ and $\hat{\hat{\lambda}}$, as well as a comparison of confidence limits, indicate there is little to choose between the two density estimators. Further, it may be stressed that the bias reduced estimators $\hat{\lambda}^*$ or $\hat{\hat{\lambda}}^*$ have been demonstrated as superior to either $\hat{\lambda}$ or $\hat{\hat{\lambda}}$ and should be used where possible.

Acknowledgement

The authors wish to thank Dr. J. Dawson for assistance with the computations.

References

Ahrens, J. H. and Dieter, U., 1974: Computer methods for sampling from gamma, beta, Poisson and binomial distributions. Computing **12**, 223–46.

Baxter, R. I., 1979: in section 4 of Food-Borne Microorganisms of Public Health Significance, Vol. 1, 3rd edition, (editors Buckle, K. A. et al.) University of New South Wales Publication Unit, Sydney.

Fisher, R. A., 1921: On the mathematical foundations of theoretical statistics. Phil. Trans. Roy. Soc. London A **222**, 309–68.

Fisher, R. A. and Yates, F., 1970: Statistical Tables for Biological, Agricultural and Medical Research (6th ed.). Oliver and Boyd, Edinburgh.

Haldane, J. B. S., 1939: Sampling errors in the determination of bacterial or virus density by the dilution method. J. Hygiene **39**, 289–293.

IMSL, International Mathematical Subroutine Library, 1980: Houston.

de Man, J. C., 1975: The probability of most probable numbers. European J. Appl. Microbiol. **1**, 67–78.

Matuszewski, T., Neyman, J. and Supinska, A., 1935: Statistical studies in questions of bacteriology. Part 1. Supplement to the J. Roy. Statist. Soc. II, 63–82.

McCrady, M. H., 1915: The numerical interpretation of fermentation tube results. J. Infectious Diseases **17**, 183–212.

Parnow, R. J., 1972: Computer program estimates bacterial densities by means of most probable numbers. Food Technology **26**, 56–62.

Salama, I. A., Koch, G. G. and Tolby, H. D., 1978: On the estimation of the most probable number of a serial dilution experiment. Commun. Statist.-Theor. Meth. **A** 7(13), 1267–81.

Manuscript received: Jan. 27, 1983

Authors' addresses:

D. J. Best,
DMS, CSIRO
P. O. Box 52,
North Ryde, NSW 2113,
Australia
J. C. W. Rayner,
Mathematics Department,
University of Otago,
P. O. Box 56,
Dunedin
New Zealand.

Biom. J. 27 (1985) 2, 173–180

A Note on the Analysis of a $p \times 2^2$ Design

B. McKee and A. M. Kshirsagar

University of Michigan, Biostatistics Dept., Ann Arbor, Mich.

Abstract

Tyagi has given a method of constructing a balanced factorial experiment of the $p\times 2^2$ type. A complete analysis of such a design with and without recovery of interblock information, using the eigenvalues and eigenvectors of the C-matrix of the design, is given in this paper. Khatry & Shah's method of producing uniformly better estimators is employed to combine the inter and intra block estimates.

Key words: Balanced asymmetrical design, Inter block estimates, Intra block estimates, Eigenvectors and values of C-matrix

1. Introduction

Tyagi (1971) has given a method of constructing a $p\times 2^2$ design which is a Balanced Asymmetrical Factorial (BAF) design. He derived the loss of information on the partially confounded interactions by the method of fitting constants. His main objective was to construct the design rather than to provide its analysis. In this note, we derive the loss of information by a simpler method, which also gives a simpler method of analysis of the design. The inter block estimates when block effects are random are also provided, and the weights for combining the intra and inter block estimates which produce uniformly better estimates as given by Khatir & Shah (1974) are also derived.

2. Incidence Matrix of the Design

Consider three factors A, B, C, with p, 2, and 2 levels respectively. The vector $\underline{t}$ of the $p\times 2^2$ treatment effects is given by

(2.1) $$\underline{t}' = [t(0,0,0), t(0,1,1,), \ldots, t(a,0,0), t(a,1,1), \ldots, t(p,0,0), t(p,1,1) \mid t(0,0,1), t(0,1,0), \ldots, t(a,0,1), t(a,1,0) \ldots, t(p,0,1), t(p,1,0)],$$

where $t(i, j, k)$ is the effect of the treatment in which A is at level i, B at level j, and C at level k. Let x_0 stand for the combinations (0, 0) & (1, 1) of B & C and x_1 stand for the combinations (0, 1) & (1, 0). Thus the first half of $\underline{t}$ consists of elements $t(a, x_0)$ and the second half consists of $t(a, x_1)$ where $a = 0, 1, 2, \ldots, p-1$. For convenience we shall also denote by t_i, the i-th element of $\underline{t}$. To construct the $p \times 2^2$ design, Tyagi starts with a design with p treatments in b^* blocks of sizes $k_1^*, k_2^*, \ldots, k_{b^*}^*$ such that each treatment is replicated r^* times and any two treatments occur together in λ^* blocks. If N^* is the $p \times b^*$ incidence matrix of this design, it is easy to see that

(2.2) $$N^* E_{b^*1} = r^* E_{pl} ,$$

(2.3) $$E_{1p} N^* = [k_1^*, \ldots, k_p^*] ,$$

where E_{1m} denotes a $1 \times m$ matrix with all unit elements. Also

(2.4) $$N^* N^{*\prime} = (r^* - \lambda^*)\, I + \lambda^* E_{pp} .$$

From (2.3) and (2.4),

(2.5) $$k_1^{*2} + \ldots + k_{b^*}^{*2} = r^* p + p\,(p-1)\,\lambda^* .$$

The BAF $p \times 2^2$ design is now constructed from this design as follows: if the a-th treatment occurs in the j-th block of this design, the j-th block of the BAF design will have the treatments (a, x_0), $(a = 0, 1, \ldots, p-1)$; otherwise it will have the treatments (a, x_1). This procedure gives b^* blocks with $k = 2p$ treatments in each block. We will number these blocks from 1 to b^*. We now add $b^* - 2r^*$ more blocks with treatments (a, x_0), if $b^* - 2r^* > 0$. If $b^* - 2r^* < 0$, we add $2r^* - b^*$ more blocks with treatments (a, x_1). In case 1 (i.e. when $b^* - 2r^* > 0$), we have in all $b = 2b^* - 2r^*$ blocks of size $k = 2p$ such that each of the $p \times 2^2$ treatments is replicated $r = b^* - r^*$ times. In case 2 (i.e. when $b^* - 2r^* < 0$), $b = 2r^*$, $k = 2p$ and $r = r^*$. From the method of construction, it is easy to verify that the incidence matrix N of this BAF design is (using $\otimes$ to denote Kronecker product of 2 matrices),

Case 1:

(2.6) $$N = \left[\begin{array}{c|c} N^* \otimes E_{21} & E_{p,b^*-2r^*} \otimes E_{21} \\ \hline (E_{pb^*} - N^*) \otimes E_{21} & 0 \end{array}\right] .$$

Case 2:

(2.7) $$\left[\begin{array}{c|c} N^* \otimes E_{21} & 0 \\ \hline (E_{pb^*} - N^*) \otimes E_{21} & E_{p,2r^*-b^*} \otimes E_{21} \end{array}\right] .$$

The rows of N correspond to the $p \times 2^2$ treatments in the order given by (2.1). If $\underline{T}$ denotes the vector of the total yields of the treatments, $\underline{B}$, the vector of the block totals, and $\underline{Q}$, the vector of the adjusted treatment totals, namely

(2.8) $$\underline{Q} = \underline{T} - \frac{1}{2p}\, N \underline{B} ,$$

the reduced normal equations for estimating treatment contrasts are

$$\underline{Q} = C\underline{t}_{\bullet},$$

where

$$(2.10) \qquad C = rI - \frac{1}{2p} NN',$$

with different values of r for cases 1 and 2. Also we shall denote by T_{aij} the total yield of the treatment (a, i, j), $(a = 0, 1, \ldots, p-1; i, j = 0, 1)$ and $T_{.ij}$ will stand for $\sum_a T_{aij}$.

3. Main Effects and Interactions

The degrees of freedom (d.f) carried by the main effects and interactions A, B, C, AB, AC, BC, ABC are respectively $(p-1)$, 1, 1, $(p-1)$, $(p-1)$, 1 and $(p-1)$. Let L be any $p \times (p-1)$ matrix of unit orthogonal contrasts i.e. a matrix satisfying

$$(3.1) \qquad E_{1p}L = 0,\ L'L = I_{p-1},\ LL' = I - \frac{1}{p} E_{pp}.$$

Define

$$(3.2) \qquad H_A = (1/2)\begin{bmatrix} L \otimes E_{21} \\ \hline L \otimes E_{21} \end{bmatrix},$$

$$(3.3) \qquad H_B = (1/2\sqrt{p})\begin{bmatrix} E_{p1} \otimes [-1, 1]' \\ \hline E_{p1} \otimes [-1, 1]' \end{bmatrix},$$

$$(3.4) \qquad H_C = (1/2\sqrt{p})\begin{bmatrix} E_{p1} \otimes [-1, 1]' \\ \hline E_{p1} \otimes [1, -1]' \end{bmatrix},$$

$$(3.5) \qquad H_{AB} = (1/2)\begin{bmatrix} L \otimes [-1, 1]' \\ \hline L \otimes [-1, 1]' \end{bmatrix},$$

$$(3.6) \qquad H_{AC} = (1/2)\begin{bmatrix} L \otimes [-1, 1]' \\ \hline L \otimes [1, -1]' \end{bmatrix},$$

$$(3.7) \qquad H_{BC} = (1/2\sqrt{p})\begin{bmatrix} E_{p1} \otimes E_{21} \\ \hline -E_{p1} \otimes E_{21} \end{bmatrix},$$

and

$$(3.8) \qquad H_{ABC} = (1/2)\begin{bmatrix} L \otimes E_{21} \\ \hline -L \otimes E_{21} \end{bmatrix}.$$

Then it can be readily seen that $H_A'\underline{t}$, $H_B'\underline{t}$, ..., $H_{ABC}'\,\underline{t}$ represent a complete set of mutually orthogonal unit contrasts belonging to A, B, ..., ABC respectively. Observe that

$$(3.9) \qquad CH_i = rH_i, \quad (i = A, B, C, AB, AC).$$

It is a well-known fact that (see for example Kshirsagar (1966)) if the unit vector $\underline{h}$ in the contrast $\underline{h}'\underline{t}$ is an eigenvector of C with eigenvalue $=r$, then it is unconfounded, its Best Linear Unbiased Estimate (BLUE)—also known as the intra block estimate is $\underline{h}'\underline{T}/r$, its variance is σ^2/r and its sum of squares (s.s.) is $(\underline{h}'\underline{T})^2/r$. Thus A, B, C, AB, AC are all unconfounded, and their BLUES, the variances of the BLUES, and s.s. can all be found as in a randomized block experiment. For example, the s.s. due to A will be (using g for grand total of yields)

$$\sum_{a=0}^{p-1} \frac{T_{a..}^2}{4r} - \frac{g^2}{4pr}, \tag{3.10}$$

the s.s. due to AB is

$$\left(\sum_{a=0}^{p-1} \sum_{i=0}^{1} \frac{T_{ai.}^2}{2r} - \frac{g^2}{4rp}\right) \quad \text{s.s. due to } A \;\&\; B, \tag{3.11}$$

and so on.

However, for BC we find that

$$CH_{BC} = \vartheta H_{BC}, \tag{3.12}$$

where,

$$\vartheta = r - \frac{1}{p^2}\left[2 \sum_{1}^{b^*} k_j^{*2} + p^2 (b^* - 3r^*)\right]. \tag{3.13}$$

The relative loss of information on BC is thus $1-\vartheta/r$ (Kshirsagar (1957)). From (2.8), (2.9) and (3.11) the BLUE of $H'_{BC}\underline{t}$ is

$$(1/\vartheta)\, H'_{BC}\underline{Q} = (1/\vartheta)\, H'_{BC}\left(\underline{T} - \frac{1}{2p} NB\right) \tag{3.14}$$

$$= (1/2\vartheta \sqrt{p})\left\{[BC] - \frac{1}{p}\sum_{j=1}^{b^*} (2k_j^* - p)\, B_j \mp \sum_{j'} B'_j\right\}$$

$$= d_{BC}/\vartheta, \quad \text{say} \tag{3.14}$$

where,

$$[BC] = T_{00} + T_{.11} - T_{10} - T_{01}, \tag{3.15}$$

and in (3.14) we take the $-$ sign for case 1 and $+$ sign for case 2 with j' running from b^*+1 to b which is different for case 1 and case 2. The variance of the BLUE of $H'_{BC}\,\underline{t}$ is

$$\sigma^2/\vartheta. \tag{3.16}$$

The s.s. due to BC is

$$(d_{BC})^2/\vartheta. \tag{3.17}$$

For the interaction ABC we first observe from (2.4) that

$$N^*N^{*\prime}L = (r^* - \lambda^*)\, L, \tag{3.18}$$

and so,

$$CH_{ABC} = \varphi H_{ABC}, \tag{3.19}$$

where

(3.20) $\varphi = r - 4\,(r^* - \lambda^*)/2p$.

The relative loss of information on any d.f. of ABC being $1 - \varphi/r$. The BLUE's of the $p-1$ contrasts $H'_{ABC}\underline{t}$ are, from (3.19), (2.8), (2.9), (2.6) or (2.7),

$$(3.21) \qquad (1/\varphi) H'_{ABC}\underline{Q} = (1/2\varphi)\,[L' \otimes E_{12} \mid -L' \otimes E_{12}] \left[\underline{T} - \frac{1}{2p} N\underline{B}\right]$$

$$(3.21) \qquad = (1/2\varphi)\,L'\,(\underline{T}_{x_0} - T_{x_1}) - (1/p\varphi)\,[L'N^* \mid O]\,\underline{B}\ ,$$

where $\underline{T}_{x_0}$ is the vector of the elements $T_{a00} + T_{a11}$ and $\underline{T}_{x_1}$ is the vector of the elements $T_{a01} + T_{a10}$, $(a = 0, 1, \ldots, p-1)$. Using $\underline{B}^*$ to denote the vector of the first b^* block totals only, (3.21) reduces to

$$(3.22) \qquad (1/\varphi)\, H'_{ABC}\underline{Q} = (1/\varphi)\, L'\underline{\xi}\ ,$$

where

$$(3.23) \qquad \underline{\xi} = (1/2)\,(T_{x_0} - T_{x_1}) - \frac{1}{p} N^* B^*\ .$$

Since the variance-covariance matrix of Q is $\sigma^2 C$, the variance-covariance matrix of the BLUE's in (3.22) is

$$(3.24) \qquad \frac{\sigma^2}{\varphi^2} H'_{ABC} C H_{ABC} = (\sigma^2/\varphi)\, I_{p-1}\ .$$

The s.s. due to the $p-1$ d.f. of the interaction ABC is therefore

$$(3.25) \qquad \left(\frac{1}{\varphi} H'_{ABC}\underline{Q}\right)' \left(\frac{1}{\xi} I_{p-1}\right)^{-1} \left(\frac{1}{\varphi} H'_{ABC}\underline{Q}\right) = \frac{1}{\varphi} \sum_{a=0}^{p-1} (\beta_a - \bar{\xi})^2\ ,$$

where ξ_a is the a-th element of $\underline{\xi}$, namely

$$(3.26) \qquad (1/2)\,(T_{a11} + T_{a00} - T_{a10} - T_{a10}) - (1/p) \sum_{j=1}^{b^*} n^*_{aj} B^*_j\ ,$$

and

$$(3.27) \qquad \bar{\xi} = \sum_a \xi_a/p\ .$$

4. Inter-Block Estimates

If the block effects are themselves normal independent with zero means and variance σ_b^2, we can find separate interblock estimates for contrasts belonging to BC and ABC. [Since A, B, C, AB, AC are unconfounded, they do not have interblock estimates.] From the theory of recovery of interblock information, it is known (KHATRI & SHAH (1974)) that the interblock estimate of a contrast $\underline{h}'\underline{t}$ is $\underline{h}'\tilde{\underline{t}}$ where $\tilde{\underline{t}}$ is a solution of

$$(4.1) \qquad NB = (NN')\,\tilde{\underline{t}}\ .$$

In particular, if $\underline{h}$ is unit and is an eigenvector of C with eigenvalue ϑ, $\underline{h}'\tilde{\underline{t}}$ is $\underline{h}'NB/2p\,(r-\vartheta)$. Therefore the interblock estimate of $H'_{BC}\underline{t}$ is

$$\frac{1}{2p\,(r-\vartheta)}\,H'_{BC}N\underline{B}\,, \tag{4.2}$$

where ϑ is given by (3.13). Using (3.14), this simplifies to

$$\frac{D_{BC}}{r-\vartheta}=\frac{1}{2\,(r-\vartheta)\sqrt{p}}\left\{\frac{1}{p}\sum_{j=1}^{b^*}(2k_j^*-p)\,B_j\pm\sum_{j'}B'_j\right\}, \tag{4.3}$$

and its variance is

$$(\sigma^2+2p\sigma_b^2)/(r-\vartheta)\,. \tag{4.4}$$

Similarly, the interblock estimates of $H'_{ABC}\underline{t}$ of the $p-1$ d.f. belonging to ABC are

$$\frac{1}{2p\,(r-\varphi)}\,H'_{ABC}N\underline{B}\,, \tag{4.5}$$

where φ is given by (3.20). This also reduces to (using (3.21),

$$\frac{1}{(r-\varphi)\,p}\,L'N^*\underline{B}^*\,. \tag{4.6}$$

The variance-covariance matrix of these interblock estimates is

$$\frac{(\sigma^2+2p\sigma_b^0)}{r-\varphi}\,I_{p-1}\,. \tag{4.7}$$

Khatri & Shah (1974) have given a method of combining the intra and interblock estimates to produce an estimate which is uniformly better than the intrablock estimate alone. We give these combined estimates in the next section.

5. Combined Inter & Intra Block Estimates for *BC* and *ABC*

Let m_1 be the d.f. (degrees of freedom) for the intrablock error s.s. For our design,

$$\begin{aligned} m_1 &= \text{total d.f.}-\text{blocks d.f.}-\text{treatments (adjusted) d.f.}\\ &=\begin{cases}(4p-2)\,(b^*-r^*)-(4p-1), & \text{for case 1}\\ (4p-2)\,r^*-(4p-1), & \text{for case 2}\,.\end{cases}\end{aligned} \tag{5.1}$$

Also,

$$\begin{aligned} S_1^2 &= \text{the intrablock error s.s.}\\ &= \text{the total s.s.}-\text{unadjusted block s.s.}\\ &\qquad -\text{s.s. due to } A,\ B,\ C,\ AB,\ AC,\ BC,\ ABC\,.\end{aligned} \tag{5.2}$$

All these s.s. are given earlier.

$$S_2^2 = \text{the interblock error s.s.}$$
$$= \text{unadjusted block s.s.} - \underline{\tilde{t}}' N \underline{B}$$

(5.3)
$$= \left(\frac{\underline{B}'\underline{B}}{2p} - \frac{g^2}{4pr}\right) - \frac{(D_{BC})^2}{r-\vartheta} - \frac{(L'N^*\underline{B}^*)'(L'N^*\underline{B}^*)}{p^2(r-\varphi)}$$
$$= \left(\frac{\underline{B}'\underline{B}}{2p} - \frac{g^2}{4pr}\right) - \frac{(D_{BC})^2}{r-\vartheta} - \frac{\sum_{a=0}^{p-1}(u_a-\bar{u})^2}{p^2(r-\varphi)},$$

where

(5.4)
$$u_a = a\text{-th element of } N^*B^* = \sum_{j=1}^{b^*} n_{aj}^* B_j^*,$$
$$\bar{u} = \sum u_a/p .$$

Also let

(5.5)
$$M = \frac{\left(\dfrac{d_{BC}}{\vartheta} - \dfrac{D_{BC}}{r-\vartheta}\right)^2}{1/(r-\vartheta)} + \frac{\sum_{a=0}^{p-1}\left\{\dfrac{1}{\varphi}(\xi_a - \bar{\xi}) - \dfrac{1}{p(r-\varphi)}(u_a - \bar{u})\right\}^2}{1/(r-\varphi)} .$$

(This is $\Sigma (x_j - y_j)^2/b_j$ of Khatri-Shah's equation (5.1).) Then the combined estimate of $H'_{BC}\underline{t}$ is

(5.6)
$$(1 - W_{BC})\left(\frac{d_{BC}}{\vartheta} - \frac{D_{BC}}{r-\vartheta}\right) + \frac{D_{BC}}{r-\vartheta},$$

where the weight W_{BC} is given by

(5.7)
$$W_{BC} = \frac{S_1^2/\vartheta}{(S_1^2/\vartheta) + \dfrac{m_1+2}{r-\vartheta}(S_2^2 + M)/(b-3)} .$$

Similarly the combined estimate of the contrasts $H'_{ABC}\underline{t}$ are

(5.8)
$$L'\left\{(1 - W_{ABC})\left(\frac{1}{\varphi}\underline{\xi} - \frac{1}{p(r-\varphi)}N^*\underline{B}^*\right) + \frac{1}{p(r-\varphi)}N^*B^*\right\},$$

where

(5.9)
$$W_{ABC} = \frac{S_1^2/\varphi}{(S_1^2/\varphi) + \dfrac{m_1+2}{(r-\varphi)(b-3)}(S_2^2 + M)} .$$

References

KHATRI, C. G. & SHAH, K. R., 1974: Estimation of location parameters from two linear models under normality. Comm. Statist. **3**, 647–664.

KHIRSAGAR, A. M., 1957: A note on the total relative loss of information in any design. Cal. Statist. Assoc. Bull. **7**, 78–81.

Kshirsagar, A. M., 1966: Balanced factorial experiments. J. Roy. Statist. Soc. B **28**, 559–562.
Tyagi, B. N., 1971: Confounded asymmetrical factorial designs. Biometrics **27**, 229–232.

Manuscript received: Sept. 28, 1982

Author's address:

Prof. A. M. Kshirsagar
The University of Michigan
School of Public Health
Dept. of Biostatistics
109 South Observatory
Ann Arbor, Michigan 48109
U.S.A.

Biom. J. 27 (1985) 2, 181–187

Problems in Estimating the Minimum Number of Genes Contributing to Quantitative Variation

O. Mayo and A. M. Hopkins

Waite Agricultural Research Institute,
University of Adelaide, Glen Osmond, South Australia

Abstract

This note emphasizes the problems inherent in gene counting methods like the Castle-Wright index and questions their utility.

1. Introduction

It is of considerable interest to know how many genes contribute to differences in quantitative traits between very distinct populations. This is the case for two reasons, one practical, one theoretical. The first is the prediction of response to selection, and of limits to selection response, where the number of genes affecting the trait is an important variable (Robertson, 1970). The theoretical interest, as emphasized by Lande (1981), is in discriminating between models of evolutionary change. On the Darwinian model large evolutionary changes are usually the result of the accumulation of many genes of individually small effect, while on certain other models, major evolutionary departures may relate to the effect of major genes, or of changes in gene regulation, usually influenced by a small number of genes. In addition, since major genes may be expected *a priori* to be unlikely to be beneficial (Fisher, 1930), and since the fixation of major genes affecting quantitative traits occurs at rates very dependent on population size (Robertson, 1956; Mayo and Hancock, 1981), chance effects may have been very important in determining the pattern of genetic variation between populations.

New gene number estimators purporting to be better than previous methods are therefore much to be welcomed. In this note, we consider Comstock and Enfield's (1981) method of gene number estimation for the case of multiplicative genetic effects, and Lande's (1981) generalization of the Castle-Wright index (Castle, 1921).

2. Estimation Procedures

(i) Comstock and Enfield (1981)

Comstock and Enfield (1981) defined the following parameters: V_{A0}, the additive genetic variance in a linkage equilibrium population with allele frequencies 0.5 for all genes heterozygous in the F_1; $\bar{G}_0$ the genotypic mean of such a population; $\bar{G}_s$ the genotypic mean of a population in which the favourable allele is homozygous for all genes heterozygous in the F_1; $\bar{F}_1$, $\bar{P}_1$ and $\bar{P}_2$ the mean performance of the F_1 and two parental lines; $H_s = \frac{\bar{F}_1}{\sqrt{\bar{P}_1 \bar{P}_2}}$ the heterosis; m_1 the effect of a genotype heterozygous for one advantageous allele; m_2 the genotype effect of a genotype homozygous for the favourable allele; and n the gene number to be estimated. Then Comstock and Enfield derived the following equations for estimating the number of genes when multiplicative genetic and additive non-genetic effects are assumed:

$$(1) \qquad \frac{m_1^n}{m_2^{\frac{n}{2}}} = H_s = \bar{F}_1 / \sqrt{\bar{P}_1 \bar{P}_2}$$

$$(2) \qquad \frac{\bar{G}_s}{\bar{G}_0} = \left[\frac{4m_2}{m_2 + 2m_1 + 1}\right]^n$$

$$(3) \qquad \frac{V_{A0}}{\bar{G}_0} = 2n \left[\frac{m_2 - 1}{m_2 + 2m_1 + 1}\right]^2$$

Comstock and Enfield introduced these equations to allow estimation without recalculation on a logarithmic scale (to convert the geometric to the additive model). They asserted that this method was simpler and that it was undesirable to make the assumption that "non-genetic, as well as genetic, effects were multiplicative." However, as emphasized by Brien and Eckert (1985), the difficulties associated with such mixed assumptions limit their use to cases in which specific evidence supporting the assumptions is available.

We have found a major problem with the model in that it is very sensitive to small changes in the parameters. We have used the IMSL subroutine ZSYSTEM (IMSL, 1980) to obtain estimates of m_1, m_2 and n, and the results have been rather unsatisfactory, as shown in Table 1. This is surprising, since ZSYSTEM is very powerful. As Table 1 shows, only in the two sets of data from the work of Comstock and Enfield were we able to derive estimates of gene numbers. In these cases, we obtained estimates the same as those of Comstock and Enfield, but small changes in the values of the sample statistics yielded very different results, as shown in Table 2, using Comstock and Enfield's analysis of *Tribolium* pupa weight as a starting point. For the other sets of data analysed, the particular parameters for which estimates were required by the method of Comstock

Table 1

Application of the method of Comstock and Enfield (1981)

Organism	Trait		$H_s = \frac{\bar{F}_1}{\sqrt{\bar{P}_1 \bar{P}_2}}$	$\frac{\bar{G}_s}{\bar{G}_0}$	$\frac{V_{A0}}{\bar{G}_0}$	m_1	m_2	n
Maize	yield	Irradiated 1969	1.19* −1.38	1.20	0.04	no solution obtained		
	yield	Control 1969	1.22* −1.45	1.28	0.04	,,	,,	,,
	yield	Irradiated 1970	1.26* −1.53	1.04	0.04	,,	,,	,,
	yield	Control 1969	1.69	1.13	0.02	,,	,,	,,
(Harris, Gardner & Compton, 1972)								
Fish	eye		1.32	3.36	0.11	,,	,,	,,
Maize	oil+		1.08	2.19	0.01	,,	,,	,,
Drosophila	head shape		1.002	1.07	0.000,07	,,	,,	,,
Human	skin colour		1.09	1.87	0.0004	,,	,,	,,
(collected by Lande, 1981)								
Tribolium	pupa weight		1.12	2.32	0.002	1.006	1.01	180 ≠
Mouse	weight gain		1.20	2.05	0.003	1.008	1.01	119 ≠
(Comstock and Enfield, 1981)								

* H_s not available; a range of values tested corresponding to given values of inbreeding depression.

+ already on geometric scale

≠ values of m_1, m_2 and n given by Comstock and Enfield

Table 2

The effect on estimates of m_1, m_2 and n of an increase of about 10% in H_s, $\bar{G}_s/\bar{G}_0$ or $V_{A0}/\bar{G}_0$

	$H_s = \frac{\bar{F}_1}{\sqrt{\bar{P}_1 \bar{P}_2}}$	$\frac{\bar{G}_s}{\bar{G}_0}$	$\frac{V_{A0}}{\bar{G}_0}$	m_1	m_2	n
Initial values*	1.12	2.3153	0.002,251	1.005,67	1.010,09	178.743
H_s increased	1.2	2.3153	0.002,251	1.005,80	1.009,72	192.696
$\bar{G}_s/\bar{G}_0$ increased	1.12	2.5	0.002,251	1.005,18	1.009,29	210.647
$V_{A0}/\bar{G}_0$ increased	1.12	2.3153	0.0025	1.006,30	1.011,21	160.974

*using unrounded $\bar{G}_s/\bar{G}_0$ and $V_{A0}/\bar{G}_0$ from Comstock and Enfield's tables

and Enfield were not always available, but if the method were robust, it should in general yield some estimates. This was not the case.

(ii) Lande (1981)

Lande (1981) introduced what appears to be a novel graphical technique for assessment of agreement of experimental data with the assumptions necessary for the Castle-Wright index, namely that all genes have equal effects, and that all gene action is additive i.e. no dominance or epistatic effects are present. This seems to require that means and variances should be additive on the same scale, so that such a scale may often not be obtainable. Lande listed six sets of data which fitted these assumptions to a greater or less extent, two on modified logarithmic scales. Four of these were analysed by the method of Comstock and Enfield, as shown in Table 1. A multiplicative model with additive non-genetic effects appears unnecessary in all cases.

Because of the relationship between range, variance and sample size for a normal distribution, HOPKINS (1974) emphasized that the Castle-Wright index will be very misleading in cases where sample sizes are very different and where the range, rather than equation (5) below is used. However, Lande took care to allow for these problems. Lande has derived an approximate variance for the estimate of gene number, and this would appear to allow assessment of the precision of estimates. Lande used the estimation equation

$$(4) \qquad n = \frac{(\bar{P}_1 - \bar{P}_2)^2}{8 V_G}$$

where

$$(5) \qquad V_G = V_{F2} - V_{F1}$$

or

$$(6) \qquad V_G = 2 V_{F2} - (V_{B1} + V_{B2})$$

then n_1 is the estimate using (4) and (5), and n_2 that using (4) and (6). The appropriate variances are given by

$$(7) \qquad V_n \cong n^2 \left(\frac{4\left(\frac{V_{P1}}{N_{P1}} + \frac{V_{P2}}{N_{P2}}\right)}{(\bar{P}_1 - \bar{P}_2)^2} + \frac{V_{VG}}{V_G^2} \right)$$

with

$$(8) \qquad V_{VG} \cong \frac{2 V_{F2}^2}{N_{F2}} + \frac{2 V_{F1}^2}{N_{F1}}$$

or

$$(9) \qquad V_{VG} \cong \frac{8 V_{F2}^2}{N_{F2}} + \frac{2 V_{B1}^2}{N_{B1}} + \frac{V_{B1}^2}{N_{B2}}$$

We have applied this method to data of STOLK, HURST and NISUK (1982) on the genetical determination of serum dopamine-β-hydroxylase level in rats, with the results shown in Table 3. Stolk *et al.* had established that, as in man, a single

Table 3

Number of genes affecting serum dopamine-β-hydroxylase level in rats (STOLK, HURST and NISUK, 1982)

	Males	Females
n_1	0.52 ±0.24	0.97 ±0.47
n_2	0.54 ±0.23	1.21 ±0.73

Table 4

Number of genes (n_l only) affecting traits in two pasture grasses, *Festuca pratensis* (SIMONSEN, 1977) and *Lolium perenne* (SIMONSEN, 1976)

	ear emergence diploid	tetraploid	yield diploid	tetraploid	height diploid	tetraploid
Festuca pratensis	2.50	0.61	11.41	1.03	2.91	2.81
Lolium perenne	3.99	1.71	7.70	7.26		

gene is the major determinant of the level of this enzyme, but with the interesting difference from the human case (WEINSHILBOUM, SCHROTT, RAYMOND, WEIDMAN and ELVEBACK, 1975) that in male rats there appeared to be dominant expression of the low enzyme level trait while in the females there appeared to be additive gene action. This is borne out in the gene number estimates of Table 3. While we are in this note concerned with the case of very large numbers of genes, this example illustrates how imprecise the estimates are, even when the mode of inheritance is simple and well understood. None of the estimates in Table 3 differs significantly from any other.

In a second application of this method, we compared the estimates of n_l for diploid and induced autotetraploid populations of two pasture grasses. On the assumptions given above for the Castle-Wright index, n from (4) will be twice the number of genes in the tetraploid (assuming tetrasomic segregation, as seems appropriate for an autotetraploid; cf QUIROS, 1982), and Table 4 shows the values obtained. In some cases, there is good agreement, in others not. Overall, we can say that the method is difficult to apply reliably. (Standard errors could not be obtained here, since not all the necessary variances were available.)

3. Discussion

Lande emphasized that the large numbers of freely segregating effective factors which he estimated to be present in a number of crosses between widely distinct populations were in good agreement with the neo-Darwinian theory that most genetic change reflects the (essentially additive) accumulation of the relatively small effects of many genes. However, as we have shown here, very little reliance may be placed on any of these gene number estimates. Accordingly, while Lande's conclusion may be correct, it is not necessarily supported by the data.

There are other methods of gene number estimation, in particular that of JINKS and TOWEY (1976), but they become very inaccurate for large gene numbers, the question of particular interest here. We conclude that gene counting methods are unlikely to be of great value in solving evolutionary problems, but that an approach based on association of quantitative gene effects with marker genes, though requiring very large numbers of observations, might be more productive. This approach, of course, has identified very many regions in plant and animal chromosomes which influence particular quantitative traits (cf. Chapter 4 of MAYO, 1980). As noted above, there is generally good agreement with Lande's view, but not on the basis of estimating minimum numbers of genes.

Acknowledgements

We thank C. J. BRIEN, S. R. ECKERT, T. W. HANCOCK and K. KAIN for assistance in the preparation of this paper. The senior author thanks the Wheat Industry Research Council for a grant for computing equipment.

Zusammenfassung

Dieser Beitrag unterstreicht die den Genzählmethoden inherenten Probleme wie den Castle-Wright Index und untersucht Fragen ihrer Nützlichkeit.

References

BRIEN, C. J. and ECKERT, S. R., 1985: On geometric estimation, the lognormal distribution and multiplicative model. Submitted for publication.

CASTLE, W. E., 1921: An improved method of estimating the number of genetic factors concerned in cases of blending inheritance. Science **54**, 223.

COMSTOCK, R. E. and ENFIELD, F. D., 1981: Gene number estimation when multiplicative genetic effects are assumed–growth in flour beetles and mice. Theor. Appl. Genet. **59**, 373–379.

FISHER, R. A., 1930: The Genetical Theory of Natural Selection. Oxford, Clarendon Press.

HARRIS, R. E., GARDNER, C. O. and COMPTON, W. A., 1972: Effects of mass selection and irradiation in corn measured by random S_1 lines and their testcrosses. Crop Science **12**, 594–598.

HOPKINS, A. M., 1974: A simulation approach to breeding for quantitative characters in a self-fertilising species. B. Ag. Sc. Honours Thesis, University of Adelaide.

IMSL, 1980: IMSL Library. Houston, IMSL Inc.

JINKS, J. L., and TOWEY, P., 1976: Estimating the number of genes in a polygenic system by genotype array. Heredity **37**, 69–81.

LANDE, R., 1981: The minimum number of genes contributing to quantitative variation between and within populations. Genetics **99**, 541–553.

MAYO, O., 1980: The Theory of Plant Breeding. Oxford, Clarendon Press.

MAYO, O. and HANCOCK, T. W., 1981: Fixation of genes having large or small effects on a trait with an intermediate optimum. Hum. Hered. **31** 286–290.

QUIROS, C. F., 1982: Tetrasomic segregation for multiple alleles in alfalfa. Genetics **101**, 117–127.

ROBERTSON, A., 1956: The effect of selection against extreme deviants based on deviation or on homozygosis. J. Genet. **54**, 236–248.

ROBERTSON, A., 1970: A theory of limits in artificial selection with many linked loci. In Biomathematics Vol. 1 Mathematical Topics in Population Genetics. (ed. K. Kojima) Berlin, Springer.

SIMONSEN, Ø., 1976: Genetic variation in diploid and autotetraploid populations of Lolium perenne L. Hereditas **84**, 133–156.

SIMONSEN, Ø., 1977: Genetic variation in diploid and autotetraploid populations of Festuca pratensis. Hereditas **85**, 1–24.

STOLK, J. M., HURST, J. H. & NISUK, B. C., 1982: Regulation and inheritance of dopamine-β-hydroxylase. Behav. Genet. **12**, 37–52.

WEINSHILBOUM, R. M., SCHROTT, H. G., RAYMOND, F. A., WEIDMAN, W. H. and ELVEBACK, L. R., 1975: Inheritance of very low serum dopamine-β-hydroxylase activity. Am. J. Hum. Genet. **27**, 573–585.

Manuscript received Feb. 12, 1982

Authors' addresses:

O. Mayo,
Biometry Section,
Waite Agricultural Research Institute,
University of Adelaide,
Glen Osmond,
South Australia 5064.

A. M. Hopkins,
Bureau of Sugar Experiment Station,
Meringa,
Queensland 4865.

Biom. J. 27 (1985) 2, 188

Book Review

BMDP – Statistical Software 1981. W. J. DIXON (chief editor), M. B. BROWN, L. ENGELMAN, J. W. FRANE, M. A. HILL, R. I. JENNRICH, J. D. TOPOREK, University of California Press, Berkeley – Los Angeles – London, 1981, 727 pp.

BMDP Biomedical Computer Programs have been developed since 1961. The present 1981 edition of the manual describes 42 programs designed for data analysis and statistical inference.

The BMDP programs comprise all essential practical methods of data analysis and statistics. They have a high level in regard to the statistical techniques, the numerical procedures, the data handling methods and the used control language. The programs have been written in FORTRAN and can be applied on IBM 360/370 computers as well as on many other computer systems.

In detail, the BMDP programs realize the following techniques:

Data description: Computation of elementary statistics and simple frequency tables for each variable and for each level of a grouping variable, if present. Comparison of groups by t test and analysis of variance. Graphical display of univariate and bivariate distributions.

Frequency tables: Computation and analysis of two-way and multiway tables. Tests of independence and measures of association and prediction. Analysis by the log-linear model.

Regression analysis: Multiple linear and polynomial regression. Nonlinear regression for some special models.

Analysis of variance: One-way and multiway analysis of variance and covariance. General mixed model analysis of variance. Multivariate analysis of variance.

Multivariate analysis: Canonical correlation analysis. Factor analysis. Discriminant analysis. Cluster analysis.

Life tables and survival analysis: Description of the distribution of survival times. Analysis of the influence of other variables.

Nonparametric analysis: Sign test and the Wilcoxon signed-rank test. Friedman's two way analysis of variance and Kendall's coefficient of concordance. Mann-Whitney rank-sum test and Kruskal-Wallis one-way analysis of variance. The Kendall and Spearman rank correlations.

Time series: Univariate and bivariate spectral analysis. Box-Jenkins time series analysis.

The BMDP programs allow great flexibility in the form and source of the data and in labeling the variables and the cases. The data can be given in the fixed or in the free format. Variable transformations and case selection are formulated by special BMDP statements. In the English-based control language of BMDP, repeated analyses of the same data can be described by few instructions.

J. LÄUTER

Biom J 27 (1985) 2, 189–197

An Application of Logistic Models to the Analysis of Ordinal Responses

SANDER GREENLAND

Division of Epidemiology, UCLA School of Public Health, Los Angeles, U.S.A.

Abstract

Logistic probability models–models linear in the log odds of the outcome event–have found extensive application in modelling of unordered categorical responses. This paper illustrates some extensions of logistic models to the modelling of probabilities of ordinal responses. The extensions arise naturally from discrete probability models for the conditional distribution of the ordinal response, as well as from linear modelling of the log odds of response. Methods of estimation and examination of fit developed for the binary logistic model extend in a straightforward manner to the ordinal models. The models and methods are illustrated in an analysis of the dependence of chronic obstructive respiratory disease prevalence on smoking and age.

Key words: Biometry; epidemiologic methods; odds ratio; statistics

An Application of Logistic Models to the Analysis of Ordinal Responses

In the study of the dependence of a response (outcome) variable on one or more independent variables, the method or model of choice is largely determined by the scale of measurement of the response variable. The scale of response may be one of three major types (1,2):

1) nominal (or qualitative), which involves no ordering of the response values (e.g., cause of death–cancer, heart disease, etc.);

2) ordinal, which involves only ranking of the response values without any specified distance between the values (e.g., clinical diagnostic category–clinically manifest disease, preclinical disease, disease absent); and

3) interval (including continuous response), which involves both ranking of values and distance between values (e.g., systolic blood pressure as measured in terms of mm Hg).

Regression and analysis-of-variance methods for nominal and interval responses have been widely disseminated (2–6), but parametric models for the analysis of ordinal responses have received much less development. As a result, it is common

This research supported by NCI grant no. R01-CA-16042

The author would like to thank Dr. JOHN A. ANDERSON for his helpful encouragement, and CONNIE BROWN for manuscript preparation.

to see ordinal responses treated as nominal (4), with consequent loss of order information. This paper discusses a general method for the modelling of ordinal responses, and illustrates the application of some specific models. The models preserve order information and can be fitted with currently available computer programs.

Ordinal Response Models Based on Discrete Distributions

Let R be the ordinal response variable, with $J+1$ ranked possible values $r_0, \ldots, r_J$, $\underline{X}=(X_1, \ldots, X_K)'$ be a column vector of K independent (regressor) variables (which usually includes a constant), and $\pi_j(\underline{x})$ be the conditional probability of response $R=r_j$ given $\underline{X}=\underline{x}=(x_1, \ldots, x_K)'$. One approach to modelling $\pi_j(\underline{x})$ is to assume that $\pi_j(\underline{x})=p(j; g(\underline{x}; \underline{\beta}))$, where $p(j; \vartheta)$ is some discrete probability function on $0, \ldots, J$ with parameter ϑ, and g is a function of $\underline{x}$ with (unknown) row vector of parameters $\underline{\beta}$ and range in the parameter space of p. Given data and a specification of the probability function p and function g, one can (in theory) estimate $\underline{\beta}$ by maximum likelihood. The problem thus hinges on how to choose p and g so that the true conditional distribution is well approximated, and the components of $\underline{\beta}$ have reasonably straightforward interpretations in terms of effects of $\underline{X}$ on response. For the initial choice graphical methods can be employed, such as histograms of R at various (possibly grouped) values of $\underline{X}$. Alternatively, one may fit a simple model which allows application of standard methods and then evaluate the adequacy of the model by both graphical and formal methods.

One of the simplest models involves taking p as the binomial distribution (7),

$$p(j; \vartheta)=\binom{J}{j}\vartheta^j(1-\vartheta)^{J-j}$$

and g as the linear-logistic function,

$$g(\underline{x}; \underline{\beta})=\frac{\exp(\underline{\beta}\underline{x})}{1+\exp(\underline{\beta}\underline{x})}.$$

The resulting model is given by

$$\pi_j(\underline{x})=\binom{J}{j}\frac{\exp(\underline{\beta}\underline{x}j)}{(1+\exp(\underline{\beta}\underline{x}))^J} \tag{1}$$

$$=\frac{\binom{J}{j}\exp(\underline{\beta}\underline{x}j)}{\sum_{l=0}^{J}\binom{J}{l}\exp(\underline{\beta}\underline{x}l)}.$$

The binomial choice arises naturally if the response value R can be thought of as the sum of J independent Bernoulli variates with common parameter ϑ. If $J=1$, model 1 reduces to the ordinary logistic model for binary responses (2,3).

Another simple model involves taking p as the truncated Poisson distribution (7),

$$p(j;\vartheta)=\frac{\vartheta^j/j!}{\sum_{l=0}^{J}\vartheta^l/l!}$$

and g as the linear-exponential function, $\vartheta=\exp(\underline{\beta x})$. The resulting model is given by

$$\pi_j(\underline{x})=\frac{\exp(\underline{\beta x}j)/j!}{\sum_{l=0}^{J}\exp(\underline{\beta x}l)/l!}. \tag{2}$$

If $J=1$ model 2 also reduces to the binary logistic model.

A more general approach involves modelling the $\pi_j(\underline{x})$ directly. For this it is often convenient to reparameterize in terms of the log-odds (logit) of r_j versus r_0, $\omega_j(\underline{x})=\log(\pi_j(\underline{x})/\pi_0(\underline{x}))$. Specifying $\omega_j(\underline{x})=a_j+\underline{\beta x}j$, $a_0\equiv 0$ yields

$$\pi_j(\underline{x})=\frac{\exp(a_j+\underline{\beta x}j)}{\sum_{l=0}^{J}\exp(a_l+\underline{\beta x}l)}. \tag{3}$$

Models 1 and 2 are special cases of model 3. The a_j are constants, usually constrained to be values of some function $h(j)$ (in model 1, $h(j)=\log\binom{J}{j}$; in model 2, $h(j)=-\log(j!)$). Depending on the form of $h(j)$, model 3 allows multimodality of the response distribution. The a_j may also be treated as unknown parameters, in which case no constant should be entered in $\underline{\beta x}$.

The simplest special case of model 3 specifies all $a_j=0$. This yields a model implying that for any $\underline{x}$ the J logit differences $\delta_j(\underline{x})\equiv\omega_{j+1}(\underline{x})-\omega_j(\underline{x})$ $(j<J)$ are equal. In this case model 3 may be rewritten $\delta_j(\underline{x})=\underline{\beta x}$.

One very general specification involves taking $\omega_j(\underline{x})=a_j+\underline{\beta x}y_j$, where y_j is a numeric scoring factor corresponding to response category r_j, and $y_J>\ldots>y_1>y_0=0$. This yields

$$\pi_j(\underline{x})=\frac{\exp(a_j+\underline{\beta x}y_j)}{\sum_{l=0}^{J}\exp(a_j+\underline{\beta x}y_j)}. \tag{4}$$

This model turns out to be a version of the "stereotype" regression model independently developed by J. A. ANDERSON (personal communication; see ref. 8). The y_j are usually specified *a priori*, though in fact they may be unknown parameters. In models 1 to 3, $y_j=j$. In the general model $\beta_k y_j$ may be interpreted as the increment in the log odds $\omega_j(\underline{x})$ per unit change in $\underline{X}$. If the y_j are treated as unknown parameters, two constraints must be applied to their values to determine location and scale of the scores; e.g., $y_1=0$, $y_2=1$ would do.

If *a priori* y_j scores for R are available, one can model the joint distribution of R and $\underline{X}$ using loglinear models for ordered categorical data (9). Model 4 (with the a_j as unknown parameters) then arises as the conditional distribution of R given $\underline{X}$ derived from the ordered loglinear model for the joint distribution of R and $\underline{X}$. (This is analogous to the relationship between logistic models and loglinear models when the response is binary (4).) Thus model 4 may be termed an "ordered logistic response" model. In fact, all the above models are logistic, in the sense that they are linear in the log odds ω_j. Other functions of the response probabilities may be linearly modelled, although the logistic is convenient because it is perhaps the simplest function that automatically constrains the predicted probabilities to lie between zero and one.

Other models for ordered logistic responses have been proposed (e.g. 3, 5, 10,11). Certain of these models assume that the discrete ordinal response has its origin in an underlying continuum (10,11), whereas the above models assume an inherently discrete response scale. If the goal of the analysis is to establish a structural model linking the response and regressor variables, choice between the models should at least partly be dictated by the nature of the response scale. On the other hand, many analyses are simply descriptive, such as smoothing of observed counts or proportions to minimize effects of random variation and provide basic tests of association. In the latter applications, model choice can be largely dictated by convenience. As noted by ANDERSON (8), inherently discrete-scale models may not only be easier to fit, but even in instances where continuous elements enter into response classification, the resulting clinical scale may be more fairly approximated by an inherently discrete response model rather than a model assuming an underlying continuum.

Standard approaches may be employed for assessment of fit of the above models. The simplest involve addition of product terms to the model and change of scale of the regressors $X_1, \ldots, X_K$. In order to detect influential or poorly fitted points, the methods discussed by COX (3) and PREGIBON (12) for the ordinary logistic model can be extended to produce residuals and other diagnostics for the models discussed here.

If a sufficiently large number of observations are available at each level of $\underline{X}$ and the y_j scores are specified *a priori*, formal tests of fit can be based on the usual test statistics for discrete data (4):

Let $n_j(\underline{x})$ = no. of responses $R = r_j$ observed at $\underline{X} = \underline{x}$,

$$n(\underline{x}) = \sum n_j(\underline{x}),$$

and $n_j(\underline{x}) = p(r_j; g(\underline{x}, \underline{\beta}))\, n(\underline{x})$ = predicted no. responses of $R = r_j$ at $\underline{X} = \underline{x}$.

Then the fit may be assessed by examination of the "χ^2 components" $\hat{e}_j(\underline{x}) = (n_j(\underline{x}) - \hat{n}_j(\underline{x}))/\sqrt{\hat{n}_j(x)}$ and by comparison of the Pearson statistic $\sum \hat{e}_j^2(\underline{x})$ against the χ^2 distribution with $NJ - P$ degrees of freedom, where N is the number of distinct values of $\underline{X}$ in the data and P is the number of independent parameters in the model. Fit may also be assessed by the likelihood ratio statistic or deviance,

$G^2 = -2\,(LL(\bar{\beta}) - LL_s)$, where $LL(\bar{\beta})$ is the log-likelihood of the fitted model and LL_s is the "saturated" log-likelihood, $LL_s = \sum_{\underset{\sim}{x}} \sum_{j} n_j(\underset{\sim}{x}) \log_e [n_j(\underset{\sim}{x})/n(\underset{\sim}{x})]$. Under the model, G^2 has asymptotically a χ^2 distribution with $NJ - P$ degrees of freedom.

Example: An Analysis of Prevalence Data

The data employed here are from the UCLA Chronic Obstructive Respiratory Disease (CORD) Lancaster survey (13), a population-based study of demographic and health characteristics as they relate to CORD prevalence. Table 1 summarizes the distribution and coding of the data on whites age 25–69. It was desired to examine prevalence of the two CORD diagnostic categories (possibly present and definitely present) as a function of age and smoking status. (Gender was also available but turned out to have no substantive or statistically significant relationship to CORD after control of age and smoking.) Possible CORD clinical diagnoses were r_0 = absent, r_1 = possibly present, and r_2 = definitely present. Although this is not an inherently discrete scale, the primary goal of the analysis was descriptive and the data provide a good example of application on the above models.

Given $n_j(\underset{\sim}{x})$ observations at level r_j of R and $\underset{\sim}{x}$ of X, the empirical logit is defined as $w_j(\underset{\sim}{x}) = \log_e(n_j(\underset{\sim}{x})/n_0(x))$ (3). As part of the analysis preliminary to model fitting, plots of the empirical logit differences $d_j(x) = w_{j+1}(x) - w_j(x) = \log\,(n_{j+1}(x)/n_j(x))$ with their estimated standard errors $\sqrt{1/n_{j+1}(x) + 1/n_j(x)}$ were made for x = age

Table 1
Summary of CORD data

Variable	Category: Name	Code	Percent with CORD: Definitely Present	Possibly Present	Total Number
Age:	25 – 29	– 4	1.0	19.4	98
	30 – 34	– 3	1.9	25.3	158
	35 – 39	– 2	2.5	19.7	198
	40 – 44	– 1	5.5	17.5	183
	45 – 49	0	4.6	15.3	176
	50 – 54	1	7.1	19.6	184
	55 – 59	2	7.8	19.6	204
	60 – 64	3	6.5	29.0	155
	65 – 69	4	7.8	22.2	90
Smoker:	Never	– 1	2 7	16.6	626
	Former	0	5.2	19.6	406
	Current	1	8.5	27.7	412
	Total		5.0	20.6	1446

χ^2 tests of association with CORD:

Age $\chi^2 = 30.3$, d. f., $= 16$, $p = .02$
Smoking $\chi^2 = 41.0$, d. f., $= 4$, $p < .001$

and x = smoking. Under models 1–3 these plots should yield parallel lines, and both lines should be straight if only first-order terms are required. The assumption of parallel and straight lines appeared to be a reasonable approximation.

Table 2 presents a summary of the initial maximum-likelihood model search procedure for the (ordered) $9 \times 3 \times 3$ table of the age-smoking-CORD data. Many other models were fit but are omitted for simplicity of presentation. From Table 2 it can be seen that the simplest model providing an adequate fit to the data was model 3 with a_1 and a_2 set to zero (model C in Table 2). This corresponds to the logit difference model

$$\delta_j(\underline{x}) = \beta_0 + \beta_1 x_1 + \beta_2 x_2, \quad j = 0, 1 ,$$

where x_2 = age and x_3 = smoking. Allowing the constant to vary (i.e., suppressing β_0 and treating the a_j as unknown parameters) did not significantly improve the fit (model D in Table 2). No improvement was obtained by adding quadratic terms for age or smoking, and fit was worsened by either log or exponential transformation of age (not shown). Addition of the product term age × smoking did significantly improve the fit at the .01 level (model E in Table 2). Using the general model 4 and allowing the y_j scores to vary did not produce a better fit.

Models A–E in Table 2 specify that the prevalence of CORD increases with age, is significantly higher among current than former smokers, and significantly

Table 2

Results of model fitting for CORD data (all fitting done by maximum likelihood).

Model type (see text)	Terms	Coefficient Estimate × 1000	Standard Error × 1000	Log Likelihood	Model d.f. and p-value
A. Binomial (model 1)*	Age	83	21	−1,003.2	51,
	Smoking	420	57		$p = .00$
B. Truncated Poisson* (model 2)	Age	75	21	− 985.6	51,
	Smoking	380	57		$p = .04$
C. Model 3 with $a_1 = a_2 = 0$*	Age	67	21	− 983.0	51,
	Smoking	341	57		$p = .10$
D. Model 3 with a_1 and a_2 estimated	Age	68	21	− 982.2	50,
	Smoking	350	57		$p = .10$
E. Model 3 with $a_1 = a_2 = 0$ and interaction*	Age	65	21	− 979.8	50,
	Smoking	338	58		$p = .20$
	Age × Smoking	62	25		
F. Unordered logistic model (see ref. 3, sec. 7.5) (all factors treated as nominal, no interactions)				− 970.9	32, $p = .16$
G. Saturated model (maximum of unconstrained likelihood)				− 950.8	0

* Constant entered but not listed

higher among former than never-smokers. Model E, however, additionally implies that the excess CORD prevalence associated with smoking increases dramatically with age. For example, according to Model E, at ages 25–29 the prevalence-odds of definite CORD among current smokers is higher than among never smokers by a multiplicative factor (odds ratio) of $\exp((.338+.062(-4))\,2)=1.2$, whereas at ages 65–69 this factor is increased to 3.2. The question naturally arises as to what extent the smoking-CORD association is a consequence of or is attenuated by the use of prevalence data rather than incidence data. If one postulates that the most likely change of smoking habit (if any occurs) after developing CORD is from current to former smoker, it would appear that the use of prevalence data should understate smoking effect among current smokers, in that prevalent cases will differentially shift from current to former smoker status. The significant increase in the smoking effect (as measured by the prevalence-odds ratio) with age may reflect a phenomenon such as a long exposure period required for smoking's full effect to become manifest. Classification errors are undoubtedly present in the CORD and smoking responses, but they are difficult to invoke as explanations for the strong pattern observed here.

The coefficients for models A–D in Table 2 are directly comparable, in that each represents the change in the logit difference δ_j per unit change in the corresponding variable. Note that these coefficients do not change dramatically with choice of model, although the log-likelihood does. The coefficients are far more sensitive to certain data subgroups: for example, by removing the most influential subgroup of the 27 possible combinations of age and smoking (age = 30 – 34, smoking = never), the coefficients ($\times 1000$) for model C of Table 2 become 90 for age and 396 for smoking, a shift of about one standard error for both coefficients. This level had 88 of the 1446 subjects, including two cases of CORD definitely present. No other subgroup deletion had as much as .50 standard error impact, but this instance points out the importance of sensitivity analysis in fitting the models discussed here.

Discussion

A simple algebraic device may be employed to estimate the parameters of the binomial model (model 1) using any standard maximum likelihood logistic regression program that allows entry of grouped data: in accord with the derivation of the model, every response is treated as the outcome of a sequence of J independent identically distributed Bernoulli trials, with the value r_j representing j "successes" and $J-j$ "failures". An individual with ordinal response r_j is thus analyzed as a group of J binary responses at that individual's covariate level. This treatment of the data yields a log-likelihood algebraically identical (up to a constant) to that obtained directly from Model 1; thus parameter estimates and standard errors obtained in this manner will be correct. The correct expected counts $\hat{n}_j(\underline{x})$ under the model may be obtained by direct computation from the data using the estimated parameters.

For the general case, models may be fit using any program that will do maximum likelihood fitting of user-specified functions, such as BMDP programs BMDP3R and BMDPAR (14), or the interactive generalized linear models program GLIM (15). If a sufficiently large number of observations are available at each possible level of the regressor $\underline{X}$, the models may be fit by the weighted least squares (WLS) method of GRIZZLE, STARMER and KOCH (16), available in SAS program FUNCAT (17). For model D of Table 2, the WLS method yielded coefficient estimates ($\times 1000$) of 62 for age and 352 for smoking. These are quite close to the maximum-likelihood estimates in Table 2 and were computationally cheaper to obtain. Unfortunately, the WLS method cannot handle zero observations at certain response-regressor combinations without recourse to *ad hoc* aids such as adding 1/2 to each count.

The models presented here provide an efficient means of analyzing the dependence of ordinal responses on an arbitrary set of independent factors. Nevertheless, the models embody rather strong assumptions, and users of these models should be sure to examine these assumptions by (a) plotting or tabulating the empirical rates and logits of the various response levels against levels of the independent variables, (b) applying formal goodness-of-fit, residual, and sensitivity analyses, and (c) fitting and comparing a variety of models.

As a final caution, it should be noted that the unconditional maximum likelihood method employed here may have poor small-sample properties when fitting the ordered logistic models. Recent work by PIKE et al. (18) for the binary logistic model indicates that the unconditional method incorporates excessive bias when the number of observations per level of the regressor vector $\underline{X}$ is under 6 or so. (For the CORD data there were at least 11 observations at each age-smoking level.) It is suggested that, before model fitting, the adequacy of sample size be assessed by checking for sparseness in a multiway cross-classification of the data based on R and $\underline{X}$.

References

1. SIEGEL, S., 1956: Nonparametric Statistics for the Behavioral Sciences. New York, Wiley.
2. KLEINBAUM D. G., KUPPER, L. L. MORGENSTERN, H., 1982: Epidemiology: Principles and Quantitative Methods. Belmont Lifetime Learning.
3. COX, D. R., 1970: Analysis of Binary Data. London, Methuen.
4. BISHOP, Y. M. M., FIENBERG, S. E., HOLLAND, P. W., 1975: Discrete Multivariate Analysis. Cambridge, MIT Press.
5. FIENBERG, S. E., 1981: The Analysis of Cross-Classified Categorical Data (2nd ed.), Cambridge, MIT Press.
6. DRAPER N. R., SMITH, H., 1981: Applied Regression Analysis (2nd ed.), New York, Wiley.
7. JOHNSON, N. L., KOTZ, S., 1969: Discrete Distributions. New York, Wiley.
8. ANDERSON, J. A.: Regression modelling for ordinal categorical variables. Technical Report No. 52, Department of Biostatistics, University of Washington, Seattle.
9. HABERMAN, S. J., 1979: Analysis of Qualitative Data: New Developments, Vol. 2. New York, Academic Press.

10. McCullagh, P., 1980: Regression models for ordinal data (with discussion). J. Roy. Statist. Soc. B **42**, 109–142.
11. Anderson, J. A., Philips, P. R., 1981: Regression, discrimination, and measurement models for ordered categorical variables. Appl. Statist. **30**, 22–31.
12. Pregibon, D., 1981: Logistic regression diagnostics. Ann. Statist. **9**, 705–24.
13. Final Report for Epidemiologic Studies of CORD. Agency No. ES-01473, National Institute of Environmental Health Sciences. Los Angeles, May 1982.
14. Dixon, W. J., Brown, M. B., 1979: BMDP Biomedical Computer Programs. Los Angeles: University of California Press.
15. Nelder, J. A., 1976: General Linear Interactive Modelling (GLIM). Oxford: Royal Statistical Society.
16. Grizzle, J. E., Starmer, C. F., Koch, G. G., 1969: Analysis of categorical data by linear models. Biometrics **25**, 489–504.
17. SAS User's Guide. Raleigh, NC: SAS Institute Inc., 1979.
18. Pike, M. C., Hill, A. P., Smith, P. G., 1980: Bias and efficiency in logistic analyses of stratified case-control studies. Int. J. Epidemiol. **9**, 89–95.

Manuscript received. Oct 3, 1983

Author's address:

Sander Greenland, M. S, Dr. P. H.
Associate Professor
Division of Epidemiology
UCLA School of Public Health
Los Angeles, CA 90024, U.S.A.

Biom. J. 27 (1985) 2, 198

Book Reviews

JANSSEN, J., J.-F. MARCOTORCHINO, J.-M. PROTH (eds.): New Trends in Data Analysis and Applications, North-Holland, Amsterdam – New York 1983. VII, 274 pp., $ 51.–; Dfl. 120.–.

This volume contains fifteen papers, held at two congresses on data analysis in 1981 and 1982 in Brussels. This has the following purposes:

1. to give new and usual methods that are easy to understand and to use
2. to show some applications in different areas.

For realizing this and to have clear contents the book is spreaded up into six chapters. In the first chapter preference and similarity analysis are considered. Here three papers are given:

P. MICHAUD: Opinions aggregation

J. WARNESSON: Use of the similarity aggregation method for the clustering of lexicographical works

L. ASFAR: An application of a similarity aggregation method to image processing in remote sensing.

Here some new theoretical results are discussed and applications are given. The second chapter deals with factor analysis and other topics. The following paper is given:

M. GUILLEVIC: Analysis and Classification of E.D.F. Consumers according to the Shape of their Load Curve

The third chapter is on discriminant analysis, multidimensional scaling and coding theory where two papers are given:

G. SAPORTA: Multidimensional data Analysis and Quantification of Categorical Variables

J. VAN RIJCKEVORSEL and J. WALTER: An Application of Two Generalizations of Non-Linear Principal Components Analysis

In the chapter 4 special clustering methods and pattern recognition are considered. Here two papers are included.

R. TRÉMOLIÈRES: The Generalized Percolation for Data Analysis and Pattern Recognition

J. AGUILAR-MARTIN: Self Learning Classification of Data for Process Monitoring

The chapter 5 contains four papers on the clustering analysis where classical approaches are discussed.

E. DIDAY: Problems of Clustering and Recent Advances

S. LEVI and M. TRANZO: Typological Variables Selection: An Application to Stock Price Comovement

G. LIBERT and M. ROUBENS: New Experimental Results in Cluster Validity of Fuzzy Clustering Algorithms

P. KORHONEN and E. BLAFIELD: A Comperative Study of the Clustering Approaches by Day and Scott and Symons

In the sixth chapter applications of data analysis in cancer prognosing are included. Here medical applications are discussed.

A. LE GO: Application of "Data Analysis" to the Evaluation of Clinical Trials

M. BUYSE: Correspondence Analysis as an Exploratory Tool in Medical Trials

J. P. NAKACHE, B. ASSELAIN, and P. POULLART: Contribution of Correspondence Analysis in Survival Trials

In this volume one finds results of various fields in data analysis. Alro the references in the papers are of interest to the readers.

H. Läuter

Biom. J 27 (1985) 2, 199–203

Combining the Pragmatic Attitude and the Play-the-Winner in Controlled Clinical Trials

J. JAMART

Laboratory for Mathematical Data Analysis,
Facultés Universitaires, Notre-Dame de la Paix,
Namur, Belgium

Various designs have been proposed for controlled clinical trials with a decisional rather than an explanatory approach. These management trials seek to determine the best treatment for an illness under circumstances as close as possible to those seen in the clinical situation (SACKETT and GENT, 1979). The designs constructed in such a decisional strategy include the so-called pragmatic attitude (SCHWARTZ and LELLOUCH, 1967) and various adaptive treatment assignment methods, including the play-the-winner rule (ZELEN, 1969). The purpose of this paper is to demonstrate the interest of combining both procedures, thereby retaining the main advantages of both methods without their in conveniences.

Characteristics of the combined design

Suppose a controlled clinical trial is planned to choose, for a particular disease, between a new treatment B and a classical therapy A for which one can estimate the success rate P_A from previous studies. If one accepts a maximal error β of non detection of a specified clinically relevant difference Δ in success rates in a two-tailed test with a significance level α, the total number of required patients using the classical methodology of null hypothesis tests is (SCHWARTZ et al., 1970)

$$N = 2n = \frac{(Z_\alpha + Z_{2\beta})^2}{(\text{arc sin} \sqrt{P_B} - \text{arc sin} \sqrt{P_A})^2} \tag{1}$$

where Z_α and $Z_{2\beta}$ are the normal deviates corresponding to values α and 2β and $P_B = P_A + \Delta$.

In a decisional problem the pragmatic formulation proposed by SCHWARTZ and LELLOUCH (1967) may be used. Its purpose is to permit the choice of a treatment, not because it is necessarily better, but because it is not comparatively inferior. With such a procedure one accepts that the choice may result from chance if both treatments are equally effective. In this design $\alpha = 1$ and $\beta = 0$, and one must minimize the risk γ of choosing one of the treatments if it is really the other which is the most successful. For a fixed γ error, this methodology will permit the acceptance of the best treatment without performing a statistical test at the end of the study if the computed number of required patients has been reached. For an equal sample size n' in both groups, we calculate (SCHWARTZ et al., 1970)

$$(2) \qquad n' = \frac{(Z_{2\gamma})^2}{2\,(\arcsin \sqrt{P_B} - \arcsin \sqrt{P_A})^2}$$

where $Z_{2\gamma}$ is the normal deviate corresponding to value 2γ.

Now suppose that the modified play-the-winner sampling (ZELEN, 1969) for assigning patients into both treatment groups is adopted. A success on a treatment generates assignment of the next subject to the same treatment while failure on a treatment generates assignment of the next patient to the alternative therapy. If n'_A and n'_B are the numbers of patients in both groups and p_A and p_B their success rates, it has been shown by ZELEN (1969) that

$$(3) \qquad \frac{n'_A}{n'_B} = \frac{(1-p_B)}{(1-p_A)}$$

Since the total number of required patients in a trial with unequal sample sizes in both groups is (SCHWARTZ et al., 1970)

$$(4) \qquad N' = \frac{n'}{2}\left(2 + \frac{n'_A}{n'_B} + \frac{n'_B}{n'_A}\right)$$

it follows from equations (2), (3) and (4) that the total number of required patients in a trial combining the pragmatic attitude and the modified play-the winner sampling is

$$(5) \qquad N' = \frac{(Z_{2\gamma})^2\,(2-p_A-p_B)^2}{4\,(\arcsin \sqrt{P_B} - \arcsin \sqrt{P_A})^2\,(1-p_A)\,(1-p_B)} \,.$$

The use of this procedure will lead to a smaller number of required patients than the classical methodology if, by solving equations (1) and (5) for p_B and assuming that for the well-known treatment the success rate p_A may be considered as equal to the a priori estimation P_A,

$$(6) \qquad p_B \in 1 - \frac{(1-P_A)\,[2\,(Z_\alpha + Z_{2\beta})^2 - (Z_{2\gamma})^2 \pm 2\,(Z_\alpha + Z_{2\beta})\sqrt{(Z_\alpha + Z_{2\beta})^2 - (Z_{2\gamma})^2}]}{(Z_{2\gamma})^2}$$

with the condition $Z_\alpha + Z_{2\beta} > Z_{2\gamma}$. In the particular case where $\alpha = \beta = \gamma$, if we let $Z_\alpha = kZ_{2\beta} = kZ_{2\gamma}$, condition (6) becomes

$$(7) \qquad p_B \in 2 - P_A - 2\,(1-P_A)\,(k+1)\,[k+1 \pm \sqrt{k\,(k+2)}]\,.$$

It may be shown that if the a priori estimated success rate of the classical treatment $P_A < 0.928$ (or 0.936 if $\alpha = \beta = \gamma \geqq 0.01$ and 0.942 if $\alpha = \beta = \gamma \geqq 0.05$), which is generally the case if we are looking for a new therapy, condition (7) becomes

$$(8) \qquad p_B < 2 - P_A - 2\,(1-P_A)\,(k+1)\,[k+1 - \sqrt{k\,(k+2)}]\,.$$

On the other hand, the ethical advantage which is the justification of the play-the-winner rule, namely a greater number of patients in the best treatment group, obviously remains unchanged. There is always a smaller number of subjects in the poorer treatment group with the suggested design than with the usual methodology since, considering A is the poorer treatment ($p_A < p_B$),

(9) $$\frac{n'_A}{n} = \frac{(2 - p_A - p_B)\,(Z_{2\gamma})^2}{2\,(1 - p_A)\,(Z_\alpha + Z_{2\beta})^2} < 1\,.$$

An example

Suppose that we wish to choose, for a particular disease, between a new treatment B and a well-known one A for which $P_A = 0.50$ with $\alpha = 0.05$, $\beta = 0.10$ and $\Delta = 0.10$. The total number of required patients using the classical methodology (equation (1)) is $N \simeq 1{,}051$. The design combining the pragmatic attitude and the modified play-the-winner sampling with $\gamma = 0.05$ will lead to a smaller number of required patients if (condition (6)) $p_B < 0.943$. Given that this limit is so high, the proposed design is probably more advantageous than the classical method as far as the required number of patients is concerned.

Discussion

The pragmatic attitude has been rarely used in controlled clinical trials till now. The absence of a statistical test at the end of a trial may seem less scientific to some investigators and this is probably one of the reasons for its neglect. Any design that does not provide a convincing test of the null hypothesis has little chance of being adopted by many research workers (Simon, 1977). But the unawareness of this method by many clinicians is another reason for its lack of utilization. The pragmatic attitude seems, however, an interesting procedure in the problems of practical decision where it is better adapted than the classical null hypothesis tests methodology (Schwartz and Lellouch, 1967; Schwartz et al., 1970). Its main advantage is to require a smaller number of subjects than with the usual methodology for equivalent α and γ errors.

The purpose of the play-the-winner rule is different. The reason of this procedure is mainly ethical in that it minimizes the number of patients which will receive the poorer treatment. Nevertheless, as other adaptive treatment assignment methods, the modified play-the-winner sampling has been seldom used in clinical trials because of several limitations (Simon, 1977). With this procedure, the response of a subject must be known before the next patient enters a trial and the population admitted throughout the study must be homogenous in characteristics that affect the response to treatment (Lellouch and Schwartz, 1971; Byar et al., 1976; Simon, 1977). Two other limitations of the play-the-winner have been emphasized in the literature. On the one hand, this method reveals the treatment assignment of a patient to the clinician before the patient enters a study (LELLOUCH and SCHWARTZ, 1971; SIMON, 1977). We consider this so-called selection bias a false problem because it may disappear if the clinician who recruits subjects for a trial is not the clinician responsible for the treatment assignment. On the other hand, the imbalance in both sample sizes caused by using adaptive allocation means that more patients must enter a study compared with designs using equal allocation. A study using an adaptive allocation method will thus require more time and increase the overall patient population not included in the trial and consequently deprived of the possibility of receiving the better treatment (SIMON et al., 1975; BYAR et al., 1976; SIMON, 1977). The ethical advantage of the play-the-winner rule, its main justification, is thus cancelled.

It is, therefore, interesting to combine this procedure with the pragmatic attitude. Indeed, this latter method allows a reduction in the total number of required subjects without changing the relative proportions of well and poorly treated patients. In addition, the play-the-winner method permits a minimization of the number of poorly treated patients while increasing the total number of subjects. By combining both procedures, the main advantages of each method are retained without suffering from some of their inconveniences since it allows a reduction in the number of poorly treated patients while also frequently reducing the total number of required subjects. Moreover, the combination of both methods seems logical because both are conceived for a decision making strategy. The absence of randomization may be criticized, but as WEINSTEIN (1974) has written, "if one recognizes that the purpose of experimentation is to aid in decision making and if one accepts the premises of decision theory, it is never optimal to randomize".

Summary

This paper suggests to combine the pragmatic attitude and the modified play-the-winner sampling in controlled clinical trials performed in a decisional strategy. It shows that this combination permits a reduction in the number of poorly treated patients while also frequently reducing the total number of required subjects. The main advantages of both methods are thus retained without suffering from some of their inconveniences.

Resumé

Cet article propose de combiner l'attitude pragmatique et le procédé du jeu en fonction du gagnant dans les essais cliniques contrôlés réalisés dans une optique décisionnelle. Il montre que cette combinaison permet de réduire le nombre de patients qui recevront le traitement le moins favorable mais également de diminuer dans la plupart des cas le nombre total de sujets nécessaire. Les avantages majeurs de chacune des deux méthodes sont donc conservés tandis que certains de leurs inconvénients sont supprimés.

References

BYAR, D. P., R. M. SIMON, W. T. FRIEDEWALD, J. J. SCHLESSELMAN, D. L. DEMETS, J. H. ELLENBERG, M. H. GAIL and J. H. WARE, 1976: Randomized clinical trials. Perspectives on some recent ideas. New Engl. J. Med. **295**, 74–80.

LELLOUCH, J. and D. SCHWARTZ, 1971: L'essai thérapeutique: Ethique individuelle ou éthique collective? Rev. Inst. Int. Stat. **39**, 127–136.

SACKETT, D. L. and M. GENT, 1979: Controversy in counting and attributing events in clinical trials. New Engl. J. Med. **301**, 1410–1412.

SCHWARTZ, D., R. FLAMANT and J. LELLOUCH, 1970: L'essai thérapeutique chez l'homme. Flammarion, Paris.

SCHWARTZ, D. and J. LELLOUCH, 1967: Explanatory and pragmatic attitudes in therapeutic trials. J. Chron. Dis. **20**, 637–648.

SIMON, R., 1977: Adaptive treatment assignment methods and clinical trials. Biometrics **33**, 743–749.

SIMON, R., G. H. WEISS and D. G. HOEL, 1975: Sequential analysis of binomial clinical trials. Biometrika **62**, 195–200.

WEINSTEIN, M. C., 1974: Allocation of subjects in medical experiments. New Engl. J. Med. **291**, 1278–1285.

ZELEN, M., 1969: Play-the-winner rule and the controlled clinical trial. J. Amer. Statist. Assoc. **64**, 131–146.

Manuscript received: March 3, 1983

Author's address:

J. JAMART, M. D.,
18, rue Jules Adant, Bte 10,
B 1950 Kraainem,
Belgium.

Biom. J. 27 (1985) 2, 205–215

The Problem of Nucleolar Number in Cell Nuclei and in Sections of Cell Nuclei

K. Voss, K. Roth

Pathologisches Institut der Humboldt-Universität Berlin

Abstract

In this paper the mean number $\bar{N}$ of nucleoli of cell nuclei and the relative frequency p_N of cell nuclei with N nucleoli are investigated. To solve the first problem, the stereological model of convex shaped nuclei and nucleoli and the assumption of randomly-isotropically distributed objects in space are used. The model is developed for non-vanishing thickness T of tissue sections. The relationship between the unknown number $\bar{N}$ of nucleoli in cell nuclei and the mean number $\bar{n}$ of nucleoli in nuclear sections determined by counting is $\bar{N} = \bar{n}\ (\bar{D} + T)/(\bar{d} + T)$. Here $\bar{D}$ and $\bar{d}$ are the mean values of caliper diameter for nuclei and for nucleoli, resp. The second problem is illustrated by means of some biomedical examples: The relative frequency p_N of cell nuclei with N nucleoli can be approximated by a generalized Poisson distribution in all investigated cases. Therefore the mean nucleolar number $\bar{N}$ is the essential parameter to describe the frequencies of cell nuclei with different numbers of nucleoli.

Key words: Cell nuclei, nucleoli, stereology, convex bodies, integral geometry, Poisson distribution

Introduction

The nucleoli of cell nuclei play an essential role in many biomedical investigations. Their size, shape and number are connected with genetical and metabolical processes of cells and tissues (Busch and Smetana 1970). Because of their small size (1 – 2 μm in diameter), we find the nucleoli at the borderline between light microscopy and electron microscopy. Therefore the section thickness T of usual specimens used in light microscopy (4–8 μm) is large in comparison to the nucleolar diameter. So we can see many unsectioned nucleoli in a tissue section. But it must be taken into account that the magnifications necessary for investigations of nucleoli yield small depths of focussing. The cell nuclei, on the other hand, have diameters of 5–15 μm, and in the most cases we find sectioned nuclei only.

There is the problem in morphometrical investigations of nucleoli to conclude from the observable parameters (e.g. nucleolar number of the sectioned nuclei) to the "true" parameters (the nucleolar number of the whole nuclei). To solve this problem, we need some assumptions about the shape of nucleoli and nuclei, about their spatial distributions, and about the relative frequencies p_N of nuclei with N nucleoli. In this paper we will discuss these assumptions. Further we will develop a stereological model and give a hypothesis about the relative frequencies p_N. The theoretical conclusions and results are illustrated by means of experimental investigations.

2. Stereological model

The sterological problems connected with investigations of nucleoli in tissue sections have been discussed by some authors already (Schleicher et al. 1975, Schleicher et al. 1976, Cruz-Orive 1976). But in those papers special shapes of objects were assumed (spheres or spheroids). We will use the more general model of convex shaped objects. The validity of this model can be proved very simple by observations of the shape of sectioned nuclei or projected nucleoli. The assumptions about the spatial distribution of nucleoli within the nuclei and of nuclei within the tissue, resp., will be the same as used by the other authors (random and isotropic distribution).

A defined type "i" of nuclei may be characterized by the integral of mean curvature M_i (resp. the caliper diameter $D_i = M_i/2\pi$), and the number of nucleoli N_i per nuclei (the nucleolar number). Further H_i is the absolute frequency of this type of nuclei within a given tissue volume V_0. Using the general formulas of integral geometry (Santalo 1976, Blaschke 1955), we obtain the expectation value Z_i of number of (whole or sectioned) nuclei in a section of thickness T as

(1) $$Z_i = \text{const} \cdot H_i \cdot (M_i + 2\pi T) .$$

We must take into account that there are nucleoli of type "k" within cell nuclei of type "i". Their characteristic geometrical parameters are the integral of mean curvature m_{ik} (resp. the caliper diameter d_{ik}), and the absolute frequency h_{ik}. In analogy to (1), we obtain

(2) $$z_{ik} = \text{const} \cdot h_{ik} \cdot (m_{ik} + 2\pi T) .$$

The constant factor in (1) and (2) is the same for all types of nuclei, all types of nucleoli, and all thicknesses T.

Calculating the sum of all z_{ik}, we obtain the number of nucleoli of type-"i"-nuclei within the slide as

(3) $$z_i = \sum_k z_{ik} = \text{const} \sum_k h_{ik} (m_{ik} + 2\pi T)$$
$$= \text{const} \sum_{k'} h_{ik'} \frac{\sum_k h_{ik} (m_{ik} + 2\pi T)}{\sum_{k'} h_{ik'}}$$
$$= \text{const} \cdot h_i \cdot (\bar{m}_i + 2\pi T) .$$

Here h_i is the absolute frequency of nucleoli of type-"i"-nuclei in the given tissue volume V_0, and $\bar{m}_i$ is the mean value of their curvature integrals. Summing over all types of nuclei, we obtain in the same manner

$$(4) \qquad Z = \sum_i Z_i = \text{const} \cdot H \cdot (\bar{M} + 2\pi T)$$

$$z = \sum_i z_i = \text{const} \cdot h \cdot (\bar{m} + 2\pi T) .$$

Here H is the absolute frequency of nuclei in the given tissue volume V_0, and h is the absolute frequency of nucleoli in that volume. Further $\bar{M}$ and $\bar{m}$ are the mean values of curvature integrals of nuclei and nucleoli, respectively. The quotient $\bar{n} = z/Z$ is the mean number of nucleoli in the (whole or sectioned) nuclei observable within the tissue slide of thickness T. Finally $\bar{N} = h/H$ is the mean value of nucleolar number in whole nuclei. Therefore we obtain our main equation

$$(5) \qquad \bar{N} = \bar{n} \, \frac{\bar{M} + 2\pi T}{\bar{m} + 2\pi T} .$$

Using the assumptions of convex shaped objects and of a random-isotropical object distribution in space, this equation is exact. Introducing the caliper diameter $\bar{D} = \bar{M}/2\pi$ resp. $\bar{d} = \bar{m}/2\pi$, it follows

$$(6) \qquad \bar{N} = \bar{n} \, \frac{\bar{D} + T}{\bar{d} + T} .$$

In the special case of spherical objects, the caliper diameter is equal to the diameter of the objects.

The equation (5) gives the possibility to determine the mean nucleolar number $\bar{N}$ from the quantities n, $\bar{M}$, $\bar{m}$, and T. We can determine the value of $\bar{n}$ by counting experiments. The determination of the thickness T is possible by different methods (Williams 1981, Wessel and Segschneider 1971, Pearse and Marks 1976). The mean curvature integrals $\bar{m}$ and $\bar{M}$ can be determined by means of a new object counting method (Wassilew and Voss 1982, Voss 1982).

3. Practical applications

To prove the theoretical model by means of a practical example, we have prepared and stained slides of rat liver tissue. The used HBFP-staining shows nuclei in a light background with dark red nucleoli (Simon et al. 1981). The slide thicknesses were 2 µm, 6 µm, and 10 µm, resp. In these slides, all of the same tissue, the relative frequencies f_n of nuclear segments with n nucleoli were determined by counting (tab. 1, fig. 1).

To apply the equation (5), we must know the mean values $\bar{M}$ and $\bar{m}$ of curvature integrals of the liver cell nuclei and the nucleoli. In a first experiment, we have supposed the nuclei and nucleoli as nearly spherical bodies, and have determined their mean diameters in the thickest slide ($T = 10$ µm). We have found $\bar{D} = 6.9$ µm, $\bar{d} = 1.7$ µm and $\bar{M} = 43.4$ µm, $\bar{m} = 10.7$ µm, resp.

Table 1

Relative frequencies f_n of nuclear segments with n nucleoli, mean nucleolar number $\bar{n}$ of segments and estimated mean nucleolar number $\bar{N}$ of nuclei for different slide thicknesses T

thickness T in μm	number of investigated nuclear segments	f_0	f_1	f_2	f_3	f_4	f_5	$\bar{n}$	$\bar{N}$
2	1163	.235	.395	.287	.073	.010		1.22	3.26
6	980	.114	.279	.403	.171	.027	.006	1.73	3.12
10	2293	.068	.192	.311	.276	.124	.029	2.28	3.49

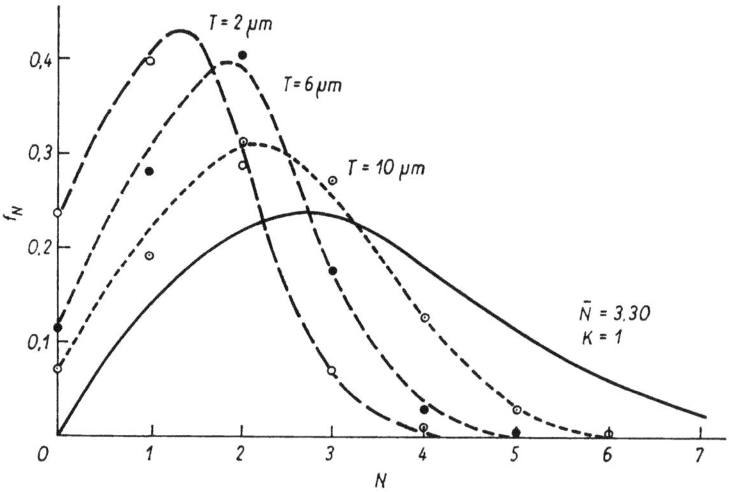

Fig. 1. Relative frequencies f_N of segments of rat liver cell nuclei with N nucleoli for different slide thicknesses T. The dashed lines are results of counts of nucleoli. The full line shows the frequency distribution of unsectioned nuclei ($T \to \infty$) calculated by means of the generalized Poisson model.

But especially the liver cell nuclei are only approximately of spherical shape, and a diameter of 7 μm is not large enough in respect to $T = 10$ μm to allow direct diameter measurements. Therefore the problem arise to determine the mean curvature integral or the mean caliper diameter for nonspherical objects in thick sections. Using the model of convex bodies and integral geometric formulas (Blaschke 1955, Santalo 1976), we can derive a relationship between the number Z of objects hitted by a slide (i.e. a large plate of area A) and the thickness T of the slide (Wassilew and Voss 1982, Voss 1982):

$$(7) \qquad Z = \varrho \frac{A}{2\pi} \bar{M} + \varrho A T .$$

Here ϱ is the numerical density of objects in the space. The first term in this equation nearly counts the objects hitted by the upper and/or lower surface of the plate. The second term counts nearly the objects within the plate. This equation (7) follows also from (4). The number of objects within the large volume $V_0 = AT$

is in proportion to ϱ. Therefore we obtain

$$(8) \qquad Z = c \cdot \varrho \cdot (\bar{M} + 2\pi T)$$

with always the same constant c. For very large thicknesses $T \to \infty$, the equation

$$(9) \qquad Z = \varrho A T$$

is valid. The comparison between (8) and (9) yields $c = A/2\pi$, and therefore equation (7) follows.

By counting the interesting objects, hitted by a field of measurement (e.g. a microscope field of view), for slides of different thicknesses, we can calculate a regression line for the count results (fig. 2). We have determined the optimal re-

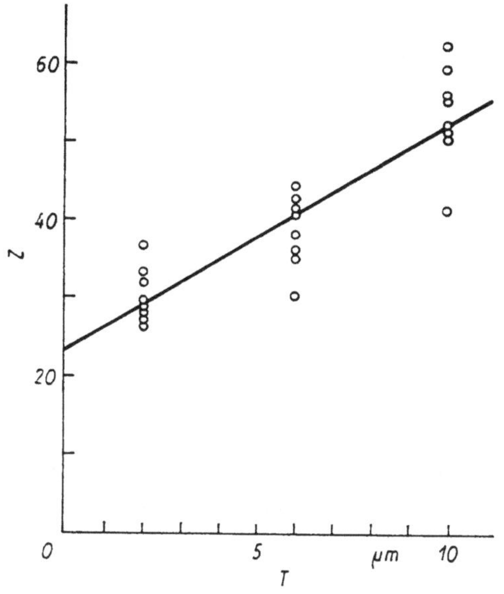

Fig. 2. Number Z of rat liver cell nuclei in a field of view, 7 000 µm² in size, for different slide thicknesses T. The regression line is $Z = 23.0 + 2.91\ T/\mu m$.

gression line $Z = a + bT$ with $a = 23.0$ and $b = 2.91\ \mu m^{-1}$. With respect to $A = 100 \times \times 70\ \mu m^2$, we obtain

$$\varrho = 0.416\ 10^{-3}\ \mu m^{-3}$$
$$\bar{M} = 49.7\ \mu m$$
$$\hat{D} = 7.9\ \mu m\ .$$

The mean value $\bar{M}$ is larger than that has been obtained by direct determination of diameters. This fact is caused by the finite thickness of $T = 10\ \mu m$ and the thereby following underestimation of the caliper diameter. Using equation (5), we obtain with $\bar{d} = 1.7\ \mu m$ and $\bar{D} = 7.9\ \mu m$ the values $\bar{N}$ of table 1. The mean result $\bar{N} = 3.30 \pm 0.20$ shows a relative standard deviation of only 6 %.

Now we must solve the problem of determination of relative frequencies p_N of nuclei with N nucleoli. To overcome this problem, we investigate count results obtained for whole nuclei. Such investigations were performed already many years ago. In fig. 3, count results for fatty cells of mouse are represented (Shea and Leb-

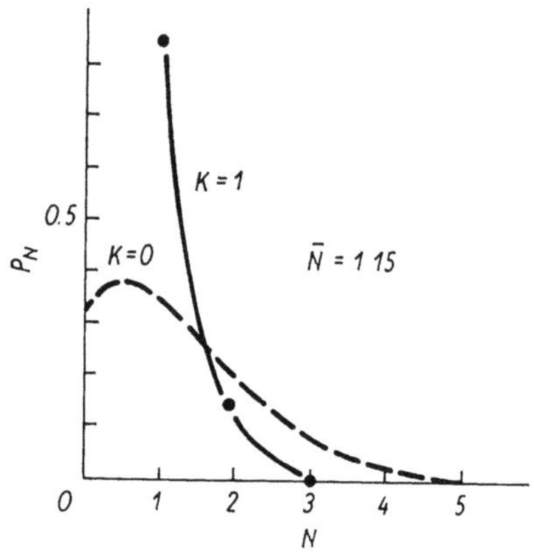

Fig. 3. Relative frequencies p_N of fatty cell nuclei of mouse with N nucleoli. The points are the experimental count results. The dashed line represents the expected values following from the usual Poisson model. The full line shows the values calculated by means of the generalized Poisson model with $K=1$.

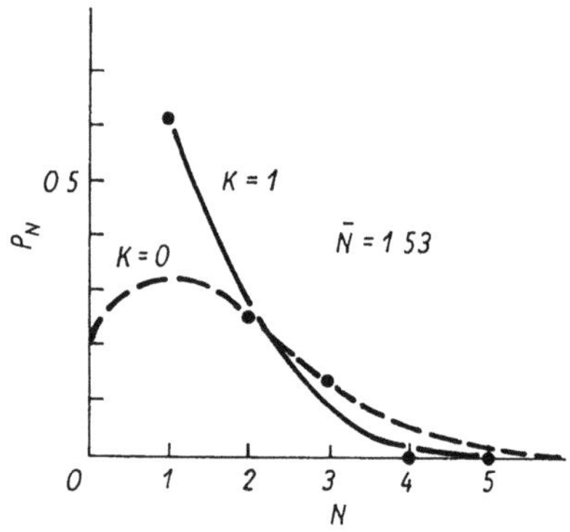

Fig. 4. Relative frequencies p_N of human oocytes with N nucleoli. Line characteristics as in fig. 3.

LOND 1966). With $\bar{N}=1.15$, the dashed line shows the expected values of p_N using the assumption of Poisson distributed nucleolar numbers. The probabilities of the Poisson distribution are

$$(10) \qquad p_N = e^{-\bar{N}} \frac{\bar{N}^N}{N!}, \qquad N=0, 1, 2 \ldots$$

In fig. 4 the relative frequencies of human oocytes with N nucleoli are represented (Stahl and Luciani 1972). Finally we show in fig. 5 the nucleolar frequencies for rat liver cells after partial hepatectomy (Mironescu and Dragomir 1966). We find in all cases that the usual Poisson model (10) is not able to describe the count results with sufficient correctness.

4. The generalized Poisson model

The main reason for the uncorrectness of the usual Poisson model is the fact that each cell nuclei owns at least one nucleolus. But $p_0 = \exp(-\bar{N})$ is nonzero. Therefore we must modify the Poisson model. We will investigate the following statistical model (Voss 1970): There are R objects (nucleoli) in B boxes (cell nuclei) with the assumptions R, $B \gg 1$ and at least K objects in each box. So we have B_N boxes, each of them with exactly N objects ($N = K$, $K + 1$, ...). It is

$$(11) \qquad \sum_{N=K}^{\infty} B_N = B \quad \text{and} \quad \sum_{N=K}^{\infty} N \cdot B_N = R = \bar{N} \cdot B .$$

Here $\bar{N}$ is the mean number of objects within the boxes. The probability to realize a special distribution (B_K, B_{K+1}, ...) is in proportion to Boltzmann's frequency

$$(12) \qquad W = \frac{R!}{\prod_{b=1}^{B} r_b!} \frac{B!}{\prod_{N=K}^{\infty} B_N!} .$$

The first term in this equation is the frequency caused by changing the objects. Here r_b are the object numbers within the B boxes. The second term is the frequency caused by changing the boxes. Like in the first term for objects, the change of two boxes with the same number N of objects gives no new distribution. Therefore the expression $B_N!$ in the nominator follows.

In the first term, the numbers r_b can have the values K, $K + 1$, ... with the frequencies B_K, B_{K+1}, ... resp. Therefore we obtain

$$(13) \qquad \prod_{b=1}^{B} r_b! = \prod_{N=K}^{\infty} (N!)^{B_N}$$

and finally

$$(14) \qquad W = \frac{R!B!}{\prod_{N=K}^{\infty} (N!)^{B_N} B_N!} .$$

This frequency W is a measure for the probability to realize the special distribution (B_K, B_{K+1}, ...). In practice, that distribution will be realized, for which W will be a maximum in respect to the conditions (11). Because of $\ln(W)$ is a monotonic function of W, we can investigate also the expression $\ln(W)$ to find the most probable distribution. We use Stirling's formula for large values B_N:

$$(15) \qquad \ln(B_N!) = B_n \cdot \ln(B_N) \cdot \left[1 - \frac{1}{\ln(B_N)} + \frac{1}{2 \cdot B_N} + \ldots\right] \approx B_N \cdot \ln(B_N) .$$

Then we obtain approximately

$$(16) \qquad \ln(W) = \ln(R!) + B \cdot \ln(B) - \sum_{N=K}^{\infty} [B_N \cdot \ln(B_N) + B_N \cdot \ln(N!)] .$$

Using the first equation (11), it follows

$$(17) \qquad \ln \frac{W}{R!} = - \sum_{N=K}^{\infty} B_N \cdot \left[\ln \frac{B_N}{B} + \ln(N!)\right]$$

or

$$(18) \qquad B^{-1} \ln \frac{W}{R!} = = \sum_{N=K}^{\infty} \frac{B_N}{B} \cdot \left[\ln \frac{B_N}{B} + \ln(N!)\right]$$

We call the expression $B^{-1}\ln(W/R!)$ the entropy S of the special distribution $(B_K, B_{K+1}, \ldots)$. Because of the monotony of the logarithmic function, the problem $W\to\max$ is the same as the problem $S\to\max$. With $p_N=B_N/B$ and (11), we can formulate this problem as

$$S=-\sum_{N=K}^{\infty} p_N\cdot[\ln(p_N)+\ln(N!)]\to\max\,, \tag{19}$$
$$\sum_{N=K}^{\infty} p_N=1 \quad\text{and}\quad \sum_{N=K}^{\infty} N\cdot p_N=\bar{N}\,.$$

The entropy expression is here another than that of Boltzmann's entropy. The reason for that is caused in the assumed model, different from the Boltzmann model (Voss 1970).

The variation problem $S\to\max$ can be solved by means of Lagrange's method:

$$S'=S-\alpha\sum_{N=K}^{\infty} p_N-\beta\sum_{N=K}^{\infty} N\cdot p_N\to\max\,. \tag{20}$$

It is then

$$\delta S'=-\sum_{N=K}^{\infty}\delta p_N\cdot[\ln(p_N)+\ln(N!)+1+\alpha+\beta_N]=0 \tag{21}$$

and for arbitrary δp_N follows

$$\ln(p_N)+\ln(N!)+1+\alpha+\beta N=0 \tag{22}$$

or

$$p_N=e^{-(1+\alpha)}\frac{e^{-\beta N}}{N!}\,. \tag{23}$$

This is the general expression for the frequencies p_N of the generalized Poisson model. With $c=\exp(-1-\alpha)$ and $a=\exp(-\beta)$ we obtain

$$p_N=c\frac{a^N}{N!}\quad\text{for}\quad N=K,\,K+1,\ldots \tag{24}$$
$$\sum_{N=K}^{\infty} p_N=1 \quad\text{and}\quad \sum_{N=K}^{\infty} N\cdot p_N=\bar{N}\,.$$

The parameters a and c must be determined in such a manner that the two conditions are fulfilled. For $K=0$, it follows with $a=\bar{N}$ and $c=\exp(-\bar{N})$ the usual Poisson model (10).

Especially for $K=1$ we obtain the conditions

$$c\sum_{N=1}^{\infty}\frac{a^N}{N!}=c\,(e^a-1)=1 \tag{25}$$
$$c\sum_{N=1}^{\infty} N\frac{a^N}{N!}=ca\cdot e^a=\bar{N}\,.$$

Therefore the transcendent equation

$$(26) \qquad a = \bar{N}\,(1 - e^{-a})$$

must be solved in order to determine the parameters a and c as functions of $\bar{N}$. This equation can be solved iteratively, beginning with an initial value a_0. Approximations for a_0 can be found in table 2.

Table 2

Parameters a and frequencies p_N for the generalized Poisson model with $K=1$ as functions of different values of the mean nucleolar number $\bar{N}$

$\bar{N}$	a	p_1	p_2	p_3	p_4	p_5	p_6	p_7
1.0	0.	1.0000						
1.1	0.194	.9063	.0878	.0057	.0003			
1.2	0.376	.8235	.1550	.0194	.0018	.0001		
1.3	0.550	.7501	.2062	.0378	.0052	.0005		
1.4	0.715	.6846	.2449	.0584	.0104	.0015	.0002	
1.5	0.874	.6258	.2735	.0797	.0174	.0030	.0004	
1.6	1.027	.5728	.2942	.1007	.0259	.0053	.0009	.0001
1.7	1.175	.5250	.3084	1208	.0355	.0083	.0016	.0003
1.8	1.318	.4816	.3175	.1395	.0460	.0121	.0027	.0005
1.9	1.458	.4422	.3223	.1566	.0571	.0166	.0040	.0008
2.0	1.594	.4064	.3238	.1720	.0685	.0218	.0058	.0013
2.5	2.232	.2685	.2995	.2228	.1243	.0555	.0206	.0066
3.0	2.821	.1786	.2519	.2369	.1671	.0943	.0443	.0179
3.5	3.381	.1190	.2012	.2268	.1917	.1296	.0730	.0353
4.0	3.921	.0793	.1555	.2032	.1992	.1562	.1021	.0572

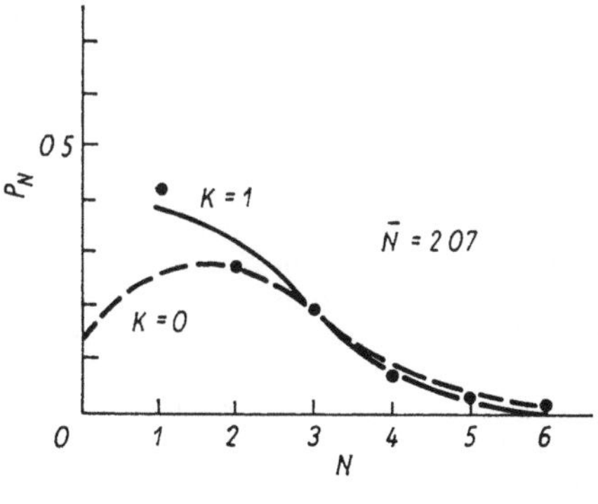

Fig. 5. Relative nucleolar frequencies p_N of rat liver cell nuclei after partial hepatectomy. Line characteristics as in fig. 3.

Using the formulas (25), we can calculate the frequencies p_N for the examples in fig. 3, 4, 5. In fig. 3 we obtain for $\bar{N}=1.15$ the value $a=0.2864$ and therefore the full line. In fig. 4 we obtain for $\bar{N}=1.53$ the value $a=0.9207$, and in fig. 5 for $\bar{N}=2.07$ the value $a=1.6868$. In all cases, we have always $p_0=0$. The generalized Poisson model with $K=1$ is able to describe the experimental results sufficiently.

Sometimes it is biologically determined that each nucleus has at least two nucleoli. Therefore the generalized Poisson model with $K=2$ must be used. An example is given in fig. 6, namely cell nuclei of a weak differentiated ovarian serous

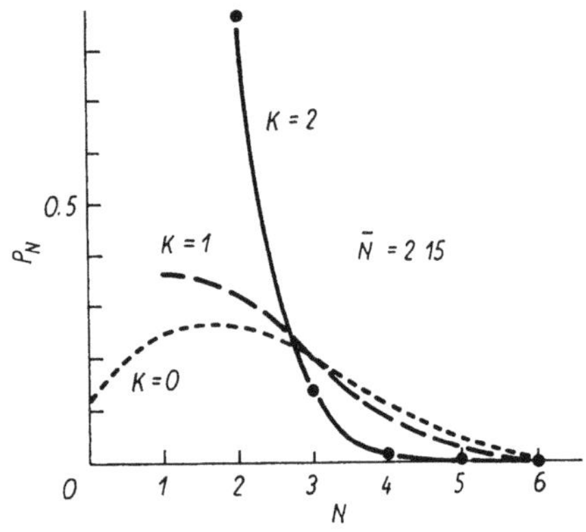

Fig. 6. Relative frequencies p_N of cell nuclei of tumour tissue, described by means of the generalized Poisson model with $K = 2$. The points are the experimental count results.

papillary tumour (Koss 1974). In this case only the model with $K = 2$ is able to describe the experimental results.

Further investigations have been made for five cell types with $K = 2$ (Koss 1974), nine cell types of mouse with $K = 1$ (Shea and Leblond 1966), nine different metabolic states of rat liver cells ($K = 1$, Mironescu and Dragomir 1966), seven different states of development of human oocytes ($K = 1$, Stahl and Luciani 1972), and for other cell types (Meinders-Groeneveld and James 1971, Petersen and Therkelsen 1962, Schleicher et al. 1976). All these investigations have been shown that the p_N can be described sufficiently by means of the generalized Poisson model with the two parameters K and $\bar{N}$.

Zusammenfassung

In dieser Arbeit werden die mittlere Nukleolenanzahl $\bar{N}$ von Zellkernen und die relative Häufigkeit p_N von Zellkernen mit N Nukleolen untersucht. Um das erste Problem zu lösen, wird ein stereologisches Modell für Zellkerne und Nukleolen konvexer Form entwickelt. Es wird vorausgesetzt, daß die Zellkerne und die Nukleolen eine zufällig homogene und isotrope räumliche Verteilung besitzen. Das Modell gilt für Schnittdicken $T > 0$. Die Beziehung zwischen der unbekannten mittleren Nukleolenanzahl $\bar{N}$ pro Zellkern und der durch Zählung ermittelten mittleren Nukleolenanzahl $\bar{n}$ pro Zellkernanschnitt ist $\bar{N} = \bar{n}\,(\bar{D} + T)/(\bar{d} + T)$. Dabei sind $\bar{D}$ und $\bar{d}$ die Mittelwerte für die mittleren Breiten der Zellkerne bzw. der Nukleolen. Das zweite Problem wird durch Ergebnisse von Nukleolenzählungen an verschiedenen Zelltypen demonstriert. Es zeigt sich, daß die relativen Häufigkeiten p_N der Zellkerne mit N Nukleolen in allen Fällen näherungsweise durch eine verallgemeinerte Poissonverteilung beschrieben werden können. Damit sind die mittlere Nukleolenanzahl $\bar{N}$ und die minimale Nukleolenanzahl K pro Zellkern die wesentlichen Parameter, um die Anzahl von Nukleolen in Zellkernen beschreiben zu können.

References

Blaschke, W., 1955: Vorlesungen über Integralgeometrie. Deutscher Verlag der Wissenschaften, Berlin.

Busch, H., K. Smetana, 1970: The Nucleolus. Academic Press, New York – London

Cruz-Orive, L. M., 1976: Correction of stereological parameters from biased samples on nucleated particle phases. In: E. E. Underwood, R. De Wit, G. A. Moore (eds.), Proc. 4th Int. Congr. Stereology, Washington, pp. 79–82.

KOSS, L. G., 1974: Morphology of Cancer Cells. In: Handbuch der allgemeinen Pathologie, Geschwülste I. Springer-Verlag, Berlin – Heidelberg – New York.

MEINDERS-GROENEVELD, J., J. JAMES, 1971: Some quantitative data regarding the nucleoli in cell nuclei from rat liver of different ploidie classes. Z. Zellforsch. **114**, 165–174.

MIRONESCU, S., C. DRAGOMIR, 1966: Nucleolar number variations in rat liver cells after partial hepatectomy. Exp. Cell Res. **43**, 217–220.

PEARSE, A. D., R. MARKS, 1976: Further studies on section thickness measurements. Histochem. J. **8**, 383–386.

PETERSEN, G. B., A. J. THERKELSEN, 1962: Number of nucleoli in female and male human cells in tissue culture. Exp. Cell Res. **28**, 590–592.

SANTALO, L. A., 1976: Integral Geometry and Geometric Probability. Addison-Wesley Publ. Comp., Reading/Mass.

SCHLEICHER, A., K. ZILLES, F. WINGERT, H. J. KRETSCHMANN, 1975: Bestimmung der Anzahl von Zellen mit mehr als einem Nukleolus im histologischen Schnittpraparat. Microsc. Acta **77**, 316–330.

SCHLEICHER, A., H. J. KRETSCHMANN, F. WINGERT, K. ZILLES, 1976: Determination of the number of cells with multiple nucleoli in histological sections. In: E. E. UNDERWOOD, R. DE WIT, G. A. MOORE (eds.), Proc. 4th Int. Conf. Stereology, Washington, pp. 371–374.

SHEA, R., C. P. LEBLOND, 1966: Number of nucleoli in various cell types of the mouse. J. Morphol. **119**, 425–434.

SIMON, H., D. KRANZ, K. VOSS, K. WENZELIDES, 1981: Zur Methode einer automatisierten Mikroskopbildanalyse an histologischen Schnitten. Zbl. allgem. Pathol. **125**, 399–405.

STAHL, A., J. M. LUCIANI, 1972: Nucleoli and chromosomes–Their relationships during the meiotic prophase of the human fetal oocyte. Humangenetik **14**, 269–284.

VOSS, K., 1970: Entropie als statistisches Strukturmaß. Wissensch. Zeitschr. TU Dresden **19**, 1415–1419.

VOSS, K., 1982: Morphometry and stereology of cellular tissues. Proc. Coll. Mathem. Morphol., Stereol., Image Analysis, Prague, Sept. 1982, pp. 137–146.

WASSILEW, G., K. VOSS, 1982: Integral geometric methods in stereology and their applications in light microscopy. Proc. 3rd Eur. Sympos. Stereol., Ljubljana, June 1981, Acta Stereol. **1**, 67–71.

WESSEL, W., I. SEGSCHNEIDER, 1971: Einfache Methode zur Schnittdickenmessung histologischer Schnitt. Z. Wiss. Mikroskopie **70**, 233–235.

WILLIAMS, M. A., 1981: Sections of determined thickness for use in stereological estimations of cells. Proc. 3rd Eur. Sympos. Stereol., Ljubljana, June 1981, Stereol. Iugosl. **3**, Suppl. 1, 369–374.

Manuscript received. March 3, 1983

Author's address:

Dr. sc. nat. K. VOSS, Dipl.-Phys. K. ROTH
Institute of Pathology
Humboldt University
DDR - 1040 Berlin
Schumannstr. 20/21

Biom. J. 27 (1985) 2, 217–223

Some Approximations of the Borel-Tanner or the Generalized Poisson Distribution

M. S. H. Khan

Department of Statistics, Al-Fateh University, Tripoli, Libya

Abstract

This paper considers some approximations for the Borel-Tanner (Generalized Poisson) sums by using (i) Gram-Charlier Poisson expansion, (ii) Mixture of two Poisson distributions, (iii) Variance stabilizing technique, and (iv) negative binomial distribution. It has been found that the approximation obtained by using the negative binomial distribution seems to be more efficient than the other approximation.

Key words: Borel-Tanner distribution, generalized Poisson distribution, and negative binomial distribution; Gram-charlier expansion; mixture of two Poisson distributions; transformation of function using variance stabilizing technique; Approximation of cumulative sums.

1. Introduction

Consider a single-server steady state queueing process with Poisson arrivals (at rate α) of customers and a constant time β required for serving each customer. Suppose there are m customers in the queue when the service starts. Then the probability that the total number $m+j$ of customers are served before the queue first vanishes is given by the Borel-Tanner distribution

(1.1) $$b_j = \lambda^j m (m+j)^{j-1} e^{-\lambda(m+j)}/j!, \quad j=0, 1, 2, \ldots, \lambda = \alpha\beta < 1 .$$

This distribution was obtained by Borel (1942) for $m=1$ and by Tanner (1953) for general m.

A variant of the distribution (1.1) as obtained by writing $m\lambda = \lambda_1$ in it is

(1.2) $$b_j = \lambda_1 (\lambda_1 + \lambda j)^{j-1} e^{-(\lambda_1 + \lambda j)}/j!, \quad j=0, 1, 2, \ldots, \lambda < 1 .$$

As the Poisson distribution with parameter λ_1 is a special case of (1.2) when $\lambda = 0$, Consul and Jain (1973) have called it a generalized Poisson distribution and studied some of its properties. The probability generating function (pgf) of (1.2) can be shown to be

(1.3) $$B(S) = \sum_{j=0}^{\infty} b_j S^j = \exp \{\lambda_1 [a_0 (s-1) + a_1 (s^2 - 1) + a_2 (s^3 - 1) + \ldots]\} ,$$

where a_j is the Borel distribution given by

(1.4) $$a_j = \lambda^j (1+j)^{j-1} e^{-\lambda(1+j)}/j!, \quad j=0, 1, 2, \ldots, \lambda<1$$

and can be obtained from (1.2) when $\lambda_1 = \lambda$.

The mean and variance of (1.2) are

(1.5) $$\mu_1' = \lambda_1/(1-\lambda), \quad \mu_2 = \lambda_1/(1-\lambda)^3.$$

The distributions (1.1) and (1.2) are becoming increasingly useful in many branches of science and, therefore, there arises a need to have tables of their cumulative probabilities and also to know of their relations with other distributions. Tables of the cumulative distribution functions of (1.1) have been given by HAIGHT and BREUER (1960) for $m=1$, $\lambda=.01(.1).62$. They have also produced tables (unpublished) for $m=1$, $\lambda=.63(.01).99$ for $j \leqq 34$. OWEN (1962) has computed values of the cumulative function for $m=1(1)5$, $\lambda=.01(.01).25$.

Differentiating (1.3) with respect to S and equating the co-efficient of S^j on both sides gives the recurrence relation in b_j's as

(1.6) $$b_{j+1}(j+1) = \lambda_1 [a_0 b_j + 2a_1 b_{j-1} + \ldots + (j+1) a_j b_0], \quad b_j \equiv 0, \quad \text{if} \quad j<0,$$

and hence b_j's can also be calculated successively by using the tables of the Borel distribution.

Like other distributions it is informative and useful to construct approximations for Borel-Tanner sums. JAIN (1974) obtained very good approximations for the Borel-Tanner sums $B_x = \sum_{j=0}^{x} b_j$ when $0<\lambda<.65$ by the negative binomial sums

(1.7) $$F_x = \sum_{j=0}^{x} \Gamma(n+j)\, p^j (1-p)^n / j!\, \Gamma(n),$$

where the parameters of the negative binomial distribution as obtained by equating the mean and variance of the two distributions are given by

$$p = \lambda (2-\lambda)$$
$$n = m(1-\lambda)/(2-\lambda) = \lambda_1 (1-\lambda)/(2-\lambda).$$

Alternatively the Camp-Paulson approximation (JOHNSON and KOTZ, 1969)

(1.8) $$B_x \simeq \frac{1}{\sqrt{2\pi}} \int_{-\infty}^{y} e^{-1/2u^2} du,$$

where

$$y = \frac{1}{3}\left[\frac{9x+8}{x+1} - \frac{(9n-1)}{n}\left\{\frac{np}{(1-p)(x+1)}\right\}^{\frac{1}{3}}\right]$$
$$\times \left[\frac{1}{n}\left\{\frac{np}{(1-p)(1+x)}\right\}^{\frac{2}{3}} - \frac{1}{x+1}\right]^{\frac{-1}{2}}$$

can also be used when $\lambda<.65$.

In this paper we consider a few approximations of the Borel-Tanner sums by using (i) Gram-Charlier Poisson expansion, (ii) mixture of two Poisson distributions and (iii) variance stabilizing technique. These approximations, however, seem to be inferior to the one obtained by using the negative binomial distribution.

2. Gram-Charlier Expansion

We first derive an expression for approximating any distribution g_j with known cumulants $a_1, a_2, a_3, \ldots$ and factorial moments $a_{(1)}, a_{(2)}, a_{(3)}, \ldots$ by any known distribution p_j with cumulants $b_1, b_2, b_3, \ldots$ and factorial moments $b_{(1)}, b_{(2)}, b_{(3)}, \ldots$, and then consider the Borel-Tanner approximation as a particular case. Our method is somewhat different than the one given in JOHNSON and KOTZ (1969).

Suppose that

$$a_i = b_i + c_i \quad \text{and} \quad a_{(i)} = b_{(i)} + c_{(i)} , \tag{2.1}$$

and write the cumulant generating function

$$a_1 t + a_2 \frac{t^2}{2!} + a_3 \frac{t^3}{3!} + \ldots = b_1 t + b_2 \frac{t^2}{2!} + \ldots + c_1 t + c_2 \frac{t^2}{t!} + c_3 \frac{t^3}{3!} + \ldots \tag{2.2}$$

From (2.2) a relation in the moment generating function of g_j and p_j can be obtained as

$$\begin{aligned} \sum_{j=0}^{\infty} g_j e^{jt} &= \left(\sum_{j=0}^{\infty} p_j e^{jt}\right)\left[1 + c_1 t + (c_2 + c_1^2)\frac{t^2}{2!} + (c_3 + 3c_1c_2 + c_1^3)\frac{t^3}{3!} + \ldots\right] \\ &= \left(\sum_{j=0}^{\infty} p_j e^{jt}\right)\left[1 - c_{(1)}(1 - e^t) + \frac{c_{(2)}}{2!}(1 - e^t)^2 - \frac{c_{(3)}}{3!}(1 - e^t)^3 + \ldots\right] \end{aligned} \tag{2.3}$$

and hence rearranging the right hand side of (2.3) gives

$$\sum_{j=0}^{\infty} g_j e^{jt} = \sum_{j=0}^{\infty} e^{jt}\left[p_j - c_{(1)} \nabla p_j + \frac{c_{(2)}}{2!} \nabla^2 p_j - \ldots\right] \tag{2.4}$$

where

$$\nabla p_j = p_j - p_{j-1}, \quad \nabla^2 p_j = \nabla p_j - \nabla p_{j-1} = p_j - 2p_{j-1} + p_{j-2} , \quad \text{etc.}$$

It, therefore, follows that

$$g_j = p_j - c_{(1)} \nabla p_j + \frac{c_{(2)}}{2!} \nabla^2 p_j - \frac{c_{(3)}}{3!} \nabla^3 p_j + \ldots \tag{2.5}$$

Further, if the distribution p_j (including the Poisson and negative binomial distributions) is such that

$$\nabla^n p_j = p_j f_n(j)$$

where $f_n(j)$ is a polynomial in j, then

$$\sum_{j=0}^{x} \nabla p_j = p_x$$

$$\sum_{j=0}^{x} \nabla^2 p_j = p_x - p_{x-1} = \nabla p_x = p_1 f_1(x)$$

and in general

$$\sum_{j=0}^{x} \nabla^n p_j = \nabla^{n-1} p_x = p_x f_{n-1}(x), \quad n=1,\,2,\,\ldots,\, f_0(x)\equiv 1\,.$$

Using these values in (2.5) finally gives the cumulative distribution function of g_j in terms of $p_{j's}$ as

$$G_x = \sum_{j=0}^{x} g_j = P_x - C_{(1)} p_x + \frac{c_{(2)}}{2!}\, p_x f_1(x) - \frac{c_{(3)}}{3!}\, p_x f_2(x) + \ldots,\; P_x = \sum_{z=0}^{x} p_j\,. \tag{2.6}$$

Poisson-Charlier approximation

For the Poisson distribution $p_x = e^{-\alpha}\alpha^x/x!$, we obtain

$$p_x f_1(x) = (1 - x/\alpha)\, p_x$$

$$p_x f_2(x) = (1 - x/\alpha)^2\, p_x$$

and in general

$$p_x f_n(x) = (1 - x/\alpha)^n p_x, \quad n=0,\,1,\,2,\,\ldots$$

and therefore (2.6) can be written as

$$G_x = P_x + p_x \sum_{i=1}^{\infty} (-)^i \frac{c_{(i)}}{i!}\left(1 - \frac{x}{\alpha}\right)^{i-1}. \tag{2.7}$$

If $g_j = \lambda_1 (\lambda_1 + \lambda j)^{j-1} e^{-(\lambda_1 + \lambda j)}/j!$ is the Borel-Tanner (generalized Poisson) distribution, then equating the means of the Borel-Tanner and the Poisson distributions in (2.7) gives

$$\begin{aligned} B_x = G_x &\simeq P_x + \frac{c_{(2)}}{2!}\, p_x f_1(x) \\ &\simeq \sum_{j=0}^{x} e^{-\alpha}\frac{\alpha^j}{j!} + \frac{1}{2}\,\frac{\lambda_1\lambda\,(2-\lambda)}{(1-\lambda)^3}\left[1 - \frac{x\,(1-\lambda)}{\lambda_1}\right]\frac{e^{-\alpha}\alpha^x}{x!}, \end{aligned} \tag{2.8}$$

where $\alpha = \lambda_1/(1-\lambda)$.

Columns (4) and (5) of Table 1 give respectively the exact and approximate values of the Borel-Tanner sums as obtained by using (2.8). The approximations are quite good provided $\lambda < .3$. It has been found that the inclusion of higher values of $c_{(i)}$ does not improve on the approximation.

Remark 1: If $g_j = \binom{n}{j} p^j (1-p)^{n-j}$ is used in (2.7), then the approximation for the binomial distribution is known as the Kolmogorov's approximation (see JOHNSON and KOTE, 1969).

Remark 2: The approximation of the Borel-Tanner sums "B_x" by the negative binomial sums "F_x" by equating the means and variances of the two distributions in the expansion (2.6) was also carried out. It was, however, noticed that the inclusion of $C_{(3)}$ and $C_{(4)}$ deteriorates the approximation. The approximation of B_x by F_x are given in column (8).

Table 1

Exact and approximate values for Borel-Tanner sums

(1) λ_1	(2) λ	(3) x	(4) Exact	(5) Approximations: Poisson Charlier using (2.8)	(6) Mixture of two Poissons using (3.1)	(7) Arc Tan Transformations using (4.3)	(8) Negative binomial using (1.7)
1	.1	0	.367879	.372092	.385639	.425861	.368550
1	.1	2	.881467	.876984	.873307	.769926	.881042
1	.1	10	.999993	1.000000	.999999	.986303	.999994
1	.3	0	.367879	.417817	.280875	.329962	.375646
1	.3	2	.801928	.753831	.850343	.625543	.797685
1	.3	10	.998331	.999989	.991948	.897412	.998651
1	.5	0	.367879	.541341	.137598	.243432	.396850
1	.5	2	.726345	.676676	.681146	.496571	.719291
1	.5	10	.979344	.999533	.999946	.771642	.979407
5	.1	0	.006738	.006385	.004009	.029149	.006800
5	.1	2	.108937	.109881	.095059	.196158	.109164
5	.1	10	.957775	.957574	.959255	.872900	.957838
5	.3	0	.006738	.003729	.000800	.017541	.007480
5	.3	2	.083466	.080573	.026844	.128926	.086150
5	.3	10	.824079	.779052	.891912	.699919	.822326
5	.5	0	.006738	.000726	.000045	.009098	.009843
5	.5	2	.064353	.030009	.002773	.075490	.074439
5	.5	10	.612005	.583040	.583245	.499514	.603862
10	.1	0	.000045	.000034	.000015	.000214	.000046
10	.1	2	.002352	.002089	.001134	.015539	.002375
10	.1	10	.460484	.462086	.450486	.484031	.460311
10	.3	0	.000045	.000005	.000000	.000280	.000056
10	.3	2	.001702	.000481	.000073	.007549	.001918
10	.3	10	.255592	.293470	.157718	.298033	.255768
10	.5	0	.000045	.000000	.000000	.000092	.000097
10	.5	2	.001239	.000012	.000000	.002979	.001974
10	.5	10	.124906	.098056	.010814	.151502	.130477

3. Approximation by Mixture of Two Poisson Distributions

For approximating the Borel-Tanner sums by the mixture of two Poisson sums, we write

$$(3.1)\qquad B_x = \sum_{j=0}^{x} \frac{\lambda_1 (\lambda_1 + \lambda j)^{j-1} e^{-(\lambda_1+\lambda j)}}{j!}$$

$$= \vartheta \sum_{j=0}^{x} \frac{e^{-\alpha_1}\alpha_1^j}{j!} + (1-\vartheta) \sum_{j=0}^{x} \frac{e^{-\alpha_2}\alpha_2^j}{j!}$$

and equate the first three factorial moments of both sides getting

$$\mu_{(1)} = \frac{\lambda_1}{1-\lambda} = \vartheta\alpha_1 + (1-\vartheta)\,\alpha_2$$

$$\mu_{(2)} = \frac{\lambda_1^2(1-\lambda) + \lambda_1\lambda(2-\lambda)}{(1-\lambda)^3} = \vartheta\alpha_1^2 + (1-\vartheta)\,\alpha_2^2$$

$$\mu_{(3)} = \frac{\lambda_1^3(1-\lambda)^2 + 3\lambda_1\lambda(1-\lambda)(2-\lambda)(\lambda_1+\lambda-1) - \lambda_1(1-\lambda)^4 + \lambda_1(1+2\lambda)}{(1-\lambda)^5}$$

$$= \vartheta\alpha_1^3 + (1-\vartheta)\,\alpha_2^3 .$$

The three parameters α_1, α_2 and ϑ in terms of $\mu_{(1)}$, $\mu_{(2)}$ and $\mu_{(3)}$ can be obtained as

$$\alpha_1 = [\mu_{(2)} - \alpha_2\mu_{(1)}]/[\mu_{(1)} - \alpha_{(2)}], \quad \vartheta = [\mu_{(1)} - \alpha_2]/[\alpha_1 - \alpha_2] .$$

and α_2 is the positive real root of the equation

$$(\mu_{(2)} - \mu_{(1)}^2)\,\alpha_2^2 + (\mu_{(1)}\mu_{(2)} - \mu_{(3)})\,\alpha_2 + (\mu_{(1)}\mu_{(3)} - \mu_{(2)}^2) = 0 .$$

Column (6) of the Table 1 gives the approximate values of the Borel-Tanner sums as obtained by using (3.1). The approximations are inferior to the Poisson and negative binomial approximations.

4. Approximation by Arc Tangent Transformation

The mean and variance of x/m, where x is a Borel-Tanner variate with $\vartheta = \lambda/(1-\lambda)$, are ϑ and $\dfrac{\vartheta(1+\vartheta)^2}{m}$ respectively. The variance stabilizing transformation to remove the parameter ϑ is therefore

$$(4.1)\qquad \int \frac{d\vartheta}{\sqrt{\vartheta}\,(1+\vartheta)} = \tan^{-1}\sqrt{\vartheta} .$$

The transformed variable $\tan^{-1}\sqrt{\dfrac{x}{m}}$ has the asymptotic mean and variance $\tan^{-1}\sqrt{\vartheta}$ and $1/2m$ respectively, when m is large. If m is sufficiently large, it can be shown that $(x - m\vartheta)/\sqrt{m\vartheta(1+\vartheta)^2}$ tends to a standard normal variate implying that $\tan^{-1}\sqrt{\frac{x}{m}}$ is also a normal variate with asymptotic mean and variance $\tan^{-1}\sqrt{\vartheta}$ and $1/2m$ respectively. The transformed variable

$$(4.2)\qquad Z=\sqrt{\frac{m+1}{2}}\left[\tan^{-1}\sqrt{\frac{x}{m+1}}+\tan^{-1}\sqrt{\frac{x+1}{m+1}}-2\tan^{-1}\sqrt{\vartheta}\right]$$

is more stable and has same approximate distribution as $\sqrt{2m}\left[\tan^{-1}\sqrt{\frac{x}{m}}-\tan^{-1}\sqrt{\vartheta}\right]$.

In column (7) we compute the Borel-Tanner sums by using the formula

$$(4.3)\qquad B_x=\frac{1}{\sqrt{2\pi}}\int_{-\infty}^{y} e^{-1/2u^2}du$$

where

$$(4.4)\qquad y=\sqrt{\frac{m+1}{2}}\left[\tan^{-1}\sqrt{\frac{x+.5}{m+1}}+\tan^{-1}\sqrt{\frac{x+1.5}{m+1}}-2\tan^{-1}\sqrt{\vartheta}\right]$$

with .5 as the continuity correction.

The approximation as given by (4.3) improves as m or λ_1 increases.

References

Borel, E., 1942: Sur l'emploi du théorème de Bernoulli pour faciliter le calcul d'un infinité de coefficients. Application au problème de l'attente a un guichet, C. r. Acad. Sci., Paris. **214**, 452–456.

Consul, P. C. and Jain, G. C., 1973: A generalization of the Poisson distribution. Technometrics **15**, 791–799.

Haight, F. A. and Breuer, M. A., 1960: The Borel-Tanner distribution. Biometrika **47**, 143–150.

Jain, G. C., 1974: A Borel-Tanner distribution and its approximation to the negative binomial distribution. Math. Operationsforsch. u. Statist. **5**, 207–212.

Johnson, N. L. and Kotz, S., 1969: Discrete distributions., Houghton Mifflin Company, Boston.

Owen, D. B., 1962: Handbook of Statistical tables., Addison-Wesley Publ. Co. Inc., Reading, Mass. – London.

Tanner, J. C., 1953: A problem of interference between two queues, Biometrika **40**, 58–69.

Manuscript received Nov. 11, 1982

Author's address:

Dr. M. S. H. Khan
Dept. of Statistics
Al-Fateh University
P. O. Box 13219
Tripoli, Libya

Biom. J. 27 (1985) 2, 225–229

Critical Values of 'Maximum Studentized Residual' Statistics in Multiple Linear Regression

S. R. PAUL

Department of Mathematics, University of Windsor, Ontario, Canada

Abstract

In multiple linear regression, test for the discordancy of a single outlier in the response variable is usually based on the 'maximum studentized residual' statistic. Exact critical values for the test statistic t are not available. Upper bounds for the critical values have been found by SRIKANTAN (1961), PRESCOTT (1975) and LUND (1975). In this note we show that all these upper bounds are algebraically equivalent.

Key words: Multiple linear regression, Discordancy test, Maximum studentized residual

1. Introduction

Test for discordancy of a single outlier in the response variable in multiple linear regression is based on the 'maximum studentized residual' statistic t. Exact critical values of t are not known. Algebraic upper bounds for the critical values are found by SRIKANTAN (1961), PRESCOTT (1975) and LUND (1975). Numerical critical values have been calculated by Monte Carlo simulations for simple linear regression by TIETJEN, MOORE and BECKMAN (1973). The purpose of this note is to show that the upper bounds for the critical values of t obtained by SRIKANTAN (1961), PRESCOTT (1975) and LUND (1975) are algebraically equivalent.

2. Test of Discordancy

The multiple linear regression model is

$$(2.1) \qquad \underline{Y} = X\underline{\beta} + \underline{\varepsilon}, \quad \varepsilon \sim \mathrm{N}(0, \sigma^2 \underline{I})$$

where $\underline{Y}$ is the $(n \times 1)$ observation vector, X is the known design matrix of $n \times k$ (assumed to be of full rank) of the observations on the independent variables, $\underline{\beta}$ is a $(k \times 1)$ vector of parameters.

When the model (2.1) is fitted the residuals can be written as

(2.2) $\underline{r} = \underline{Y} - X\hat{\underline{\beta}} = (I - H)\underline{Y}$,

with

(2.3) $\text{var}(\underline{r}) = (I - H)\,\sigma^2$,

where $\hat{\underline{\beta}}$ is the least squares estimate of $\underline{\beta}$, I is an $(n \times n)$ identity matrix and $H = X(X'X)^{-1}X'$ is an idempotent matrix. From equation (2.3) we can see that the variance of the jth residual is

$$\text{var}(r_j) = (1 - x_j'(X'X)^{-1}x_j)\,\sigma^2 = (1 - h_j)\,\sigma^2,$$

where x_j' is the jth row of X and h_j is the jth diagonal element of H. If s^2 is the residual mean squared error estimate of σ^2 the jth studentized residual is

$$t_j = \frac{r_j}{s\sqrt{1-h_j}} = \frac{y_j - \hat{y}_j}{s\sqrt{1-h_j}}$$

where $\hat{y}_j$ is the predicted value of the jth observation. These residuals are used as a basis for detecting outliers in the response variable. When $t = \max\{t_j\}$ is sufficiently large, we declare the observation yielding $\max\{t_j\}$ to be a discordant outlier.

3. Critical Values of t

The studentized residuals t_j are not independent and the distribution of t is intractable and exact critical values of t can not be found. Upper bounds of the critical values are found by SRIKANTAN (1961), Prescott (1975, 1976) and LUND (1975) and approximate critical values, by simulations, are found by TIETJEN, MOORE and BECKMAN (1973).

SRIKANTAN (1961) uses the first Bonferroni inequality to derive upper bounds for the 5 percent and 1 percent critical values of the test statistic t based on the Beta distribution. The null hypothesis of no outliers is $\text{E}(\underline{Y}) = X\underline{\beta}$ and the mean-slippage alternative due to SRIKANTAN of a single outlier is $\text{E}(\underline{Y}) = X\underline{\beta} + \underline{a}$, where $\underline{a}$ is an $(n \times 1)$ vector of zeros apart from a possible non-zero value in the jth position. Define $d_j = t_j^2/(n-k)$ and $d = \max\{d_j\}$. It can be shown that the maximum likelihood ratio test rejects H_0 in favour of H_A for large values of d_j (GENTLEMAN and WILK, 1975). Thus a test based on d would be to reject the null hypothesis of no outliers in favour of the two-sided alternative $a \neq 0$ (a is the non-zero element of $\underline{a}$) if

(2.4) $d > d_\beta$

where d_β is the critical value of the distribution of d at the $\beta\,\%$ level of significance. It can be shown that d_j is distributed as beta $\left(\frac{1}{2}, \frac{n-k-1}{2}\right)$, (SRIKANTAN, 1961).

The d_j's are correlated random variables. Thus, using first-order Bonferroni inequality for the probability of n correlated random variables, the upper bound for the critical value d_0 of d_β is obtained by solving

$$(2.5)\qquad B_{d_0}\left(\frac{1}{2}, \frac{n-k-1}{2}\right) = 1-\beta/n$$

where $B_x(p, q) = \int_0^x s^{p-1}(1-s)^{q-1}/B(p, q)$ is the incomplete beta function ratio.

TIETJEN, MOORE and BECKMAN (1973) conducted a large scale simulation study to obtain critical values of t in simple linear regression. For each sample size n, M samples were generated and a straight line fitted to each sample. t was calculated from each sample and the M values of t arranged in ascending order. The percentage points of t were then read from the empirical sampling distribution. They tabulated critical values of t for $\beta = 0.1$, 0.05, 0.01 and $n = 4(1)12(2)20$, $24(6)36(12)$, 60 and 100 and the number of samples 40,000 ($n > 30$) 10,000 ($n \geqq 30$). They confirm empirically the insensitivity of the critical values to the differing variances s_j^2. PRESCOTT (1975) agrees with this insensitivity and suggests replacing s_j in the studentized residual by $\bar{s}$, where $\bar{s}^2$ is the average variance $\sum_{i-1}^{n} r_i^2/n$ introduced by BEHNKER and DRAPER (1972). Thus the approximate maximum studentized residual is:

$$t^* = \max_j |r_j/\bar{s}| = n^{\frac{1}{2}} \max_j |r_j/\sqrt{\sum_{i=1}^{n} r_i^2}| = n^{\frac{1}{2}} \max_j |e_j| .$$

STEFANSKY (1971 and 1972) developed bounds for

$$e = \max_j |e_j| .$$

The upper bound for the $(1-\beta)$ 100 % point is given as

$$[(n-k)\, F_0/n\, (n-k-i+F_0)]^{\frac{1}{2}}$$

where $F_0 = F_{1,(n-k-1)};\ (1-\beta/n)$. t^* is therefore bounded by

$$(2.6)\qquad t_0^* = [(n-k)\, F_0/(n-k-1+F_0)]^{\frac{1}{2}}.$$

PRESCOTT compared values obtained from (2.6) with the critical values of t obtained by TIETJEN, MOORE and BECKMAN (1973) for $\beta = 0.1$, 0.05 and 0.01 and $n = 4(1)12(2)20$, 24(6), 36(12), 60 and 100 and found that these values were very much identical with those of TIETJEN, MOORE and BECKMAN. Because of the connection between the Beta and the F-distribution, it can be shown that t_0^* and d_0 are connected by

$$(2.7)\qquad t_0^* = [(n-k)\, d_0]^{\frac{1}{2}}.$$

Define $\xi_j = t_j/(n-k)^{\frac{1}{2}}$. ELLENBURG (1973) showed that the joint distribution of ξ_j's is a multivariate Inverted Students function and the marginal probability density function for any ξ_j (univariate Inverted Students function) is (using LUND's (1975) notation)

$$f(\xi) = C\,(1-\xi^2)^{\frac{n-k-3}{2}} \quad \text{for} \quad \xi^2 \leq 1$$

where

$$C = \Gamma\left(\frac{n-k}{2}\right) \Big/ \pi^{\frac{1}{2}}\,\Gamma\left(\frac{n-k-1}{2}\right).$$

Using Ellenburg's results, LUND (1975) shows that the upper bound $t_0 = \xi_0\,(n-k)^{\frac{1}{2}}$ of the critical value of t can be obtained from the equation

$$\frac{1}{2}\,(1-\beta/n) = \int_0^{\xi_0} f(\xi)\,d\xi \tag{2.8}$$

solving equation (2.8) numerically he gives the most comprehensive tabulation of upper bounds for $\alpha = 0.1$, 0.05 and 0.01 and $n = 1(1)10(2)20(5)50(10)100$ and $k = 1(1)6(2)10$, 15, 25. The 5 and 1 percent values are reproduced by BARNETT and LEWIS (1978, pp. 335–36). For $k=2$ his table entries agreed with those of PRESCOTT. However, we note that, the upper bounds t_0 and t_0^* are, in general, exactly the same, since

$$\int_0^{\xi_0} C\,(1-\xi^2)^{\frac{n-k-3}{3}}\,d\xi = \frac{1}{2}\,(1-\beta/n) \Rightarrow \int_0^{\xi_0^2} C x^{-\frac{1}{2}}\,(1-x)^{\frac{n-k-3}{2}}\,dx = (1-\beta/n)$$

$$\Rightarrow \xi_0^2 = d_0, \quad \text{(where } d_0 \text{ is obtained from 2.5)}$$

$$\Rightarrow \xi_0 = d_0^{\frac{1}{2}} \Rightarrow t_0 = \xi_0\,(n-k)^{\frac{1}{2}} = \{(n-k)\,d_0\}^{\frac{1}{2}} = t_0^*\,.$$

This shows that the distributions of t and t^* are both bounded by the same distribution and confirms with the empirical investigation of TIETJEN, MOORE and BECKMAN (1973) the insensitivity of the critical values to the differing variances s_j^2. Thus if the investigator needs any critical value of the maximized studentized residual other than those tabulated in Barnett and Lewis, he can calculate t_0 via (2.5) by using the computer algorithms of CRAN, MARTIN and THOMAS (1977) or, as mentioned by PRESCOTT (1976), directly by using the necessary percentage points of F which may be read directly from the monograph of student's t (NELSON, 1975).

Acknowledgement

This research was partially supported by the Natural Sciences and Engineering Research Council of Canada.

References

Barnett, V. and Lewis, T., 1978: Outliers in Statistical Data, New York, Wiley.

Behnken, D. W. and Draper, N. R., 1972: Residuals and their variance patterns. Technometrics **14**, 101–111.

Cran, G. W., Martin, K. J. and Thomas, G. E., 1977: Remark As R 19 and algorithm AS 109. A remark on Algorithms, As 63: The Incomplete Beta Integral, As 64: Inverse of the Incomplete Beta Function Ratio. Appl. Statist. **26**, 111–114.

Ellenburg, J. H., 1973: The joint distribution of the standardized least square residuals from a general linear model. J. Amer. Statist. Assoc. **68**, 941–943.

Gentleman, J. F. and Wilk, M. B., 1975: Detecting outliers II. Supplementing the direct analysis of residuals. Biometrics **31**, 387–410.

Lund, R. E., 1975: Tables for an approximate test for outliners in linear models. Technometric **17**, 473–476.

Nelson, Lloyd, S., 1975: A Nomograph of Students *t*. J. Quality Technology **7**, 201–202.

Prescott, P., 1975: An approximate test for outliers in linear models. Technometrics **17**, 127–132.

Prescott, P., 1975: Letter to the Editor, Technometrics **18**, 125.

Srikantan, K. S., 1961: Testing for the single outlier in a regression model. Sankhya, A. **23**, 251–260.

Stefansky, W., 1971: Rejecting outliers by maximum normal residuals. Ann. Math. Statist. **42**, 35–45.

Stefansky, W., 1972: Rejecting outliers in factorial designs. Technometrics **14**, 469–479.

Tietjen, G. L., Moore, R. H. and Beckman, R. J., 1973: Testing for a single outlier in simple linear regression. Technometrics **15**, 717–721.

Manuscript received Nov. 25, 1982

Author's address:

Dr. S. R. Paul
University of Windsor
Dept. of Mathematics
401, Sunset Avenue
Windsor, Ontario
Canada

Biom. J 27 (1985) 2, 231–235

The Power Function of the Test for 'No Three Factor Interaction' in $2\times2\times2$ Contingency Tables

M. Haber

Dept. of Mathematical Sciences, Memphis State University, U.S.A.

Abstract

Three approximations to the power function of the chi-square test for the hypotheses of 'no three factor interaction' in a $2\times2\times2$ contingency table are introduced and compared. The first method is based on the sampling distribution of the logarithm of the odds ratio, the second–on the non-central χ^2 distribution and the third–on the conditional distribution of a cell entry. The last method is found to provide the closest approximation under various alternative hypotheses and sample sizes.

Key words: Chi-square, Non-central chi-square, Odds ratio, Three-dimensional contingency tables.

1. Introduction

Large-sample tests for the hypothesis stating that the three factor interaction in a multidimensional contingency table is zero are well known. The ordinary X^2 test (or LR test) requires the calculation of expected cell frequencies. Even in the simplest case, where each of the three factors has only two categories, the expected frequencies cannot be expressed in a closed form. They can, however, be calculated by solving a cubic equation (Bartlett, 1935) or through other iterative procedures, e.g. the proportional fitting algorithm (Fienberg, 1980). Goodman (1964) devised simple tests which do not require the calculation of expected frequencies and can be carried out–at least in the $2\times2\times2$ case–using only a desk-calculator.

The aim of the present work is to introduce and compare three methods for approximating the large-sample power function of the X^2 test for the hypothesis of 'no three factor interaction' in $2\times2\times2$ contingency tables with fixed rows $\times$ $\times$ columns marginal totals. Since the LR and Goodman's tests are asymptotically equivalent to the X^2 test, our results may apply these tests as well.

2. Notations

We consider a $2 \times 2 \times 2$ contingency table with fixed rows $\times$ columns totals. Such a table may arise when a binary response variable is measured on individuals classified w.r.t two factors, each with two categories. The layers of the table correspond to the "success" and "failure" responses. A sample of n_{ij} $(i, j=1, 2)$ cases is drawn from the (i, j) combination of the two factors, and the number of "successes" is denoted by x_{ij1}. We assume that $x_{ij1} \sim \text{Bin}\,(n_{ij}, p_{ij1})$ and that the four samples are independent. Denoting $p_{ij2}=1-p_{ij1}$, the hypothesis of lack of three factor interaction can be stated as:

(1) $$H_0\colon p_{111}p_{122}p_{212}p_{221}=p_{112}p_{121}p_{211}p_{222}\,.$$

Let Ψ_i $(i=1, 2)$ denote the odds ratio in the i-th row: $\Psi_i=p_{i11}p_{i22}/p_{i12}p_{i21}$. The interaction effect can be defined as $\vartheta=\Psi_1/\Psi_2$ and the null hypothesis (1) can be rewritten as $\text{H}_0\colon \vartheta=1$. A sample estimate of ϑ is

(2) $$\hat{\vartheta}=\frac{x_{111}x_{122}x_{212}x_{221}}{x_{112}x_{121}x_{211}x_{222}}\,,$$

where $x_{ij2}=n_{ij}-x_{ij1}$.

3. Approximations to the Power Function

The power of the test for H_0 is a function of the four success probabilities p_{ij1} $(i, j= =1, 2)$. We now present three approximations to this function.

(i) The log-odds-ratio (LOR) method

Let $\omega=\log\,(\vartheta)$, $\hat{\omega}=\log\,(\hat{\vartheta})$. The large-sample distribution of $\hat{\omega}$ is $N(\omega, \sigma^2)$, where

(3) $$\sigma^2=\sum_{i=1}^{2}\sum_{j=1}^{2}(n_{ij}p_{ij1}p_{ij2})^{-1}$$

(GOODMAN, 1964). Hence, the approximated power of the test of $H_0\colon \omega=0$, $H_1\colon \omega \neq 0$ is:

(4) $$\tau_1=2-\Phi\,(z^{\alpha/2}+\omega/\sigma)-\Phi\,(z^{\alpha/2}-\omega/\sigma)$$

where Φ and $z^{\alpha/2}$ are, respectively, the c.d.f and the upper $\alpha/2$ quantile of the $N(0, 1)$ distribution

(ii) The non-central χ^2 (NCχ^2) method

Given a specific alternative $\{p_{ij1}\}$ $(i, j=1, 2)$, the expected cell frequencies are $e_{ijk}=n_{ij}p_{ijk}$. Let $\{e^0_{ijk}\}$ denote the expected frequencies which are obtained by applying the proportional fitting algorithm to the "contingency table" $\{e_{ijk}\}$. Using a general result by HABER (1981), it can be shown that the distribution of the X^2 statistic under the alternative $\{p_{ijk}\}$ is asymptotically $\chi^2_{(1,\lambda)}$ where

(5) $$\lambda=\sum_{ijk}(e_{ijk}-e^0_{ijk})^2/e^0_{ijk}\,.$$

Thus, the power can be approximated by: $\tau_2 = P(\chi^2_{(1,\lambda)} >$ upper α quantile of the $\chi^2_{(1)}$ distribution).

(iii) A method based on the asymptotic conditional distribution). (ACD).

Let $\{e_{ijk}\}$ and $\{e^0_{ijk}\}$ be defined as above, and denote $p^0_{ijk} = e^0_{ijk}/n_{ij}$. Using the results of GODAMBE and HARKNESS (1975), it can be shown that the conditional distribution of x_{111} given all the two-dimensional marginal totals (of the table $\{e_{ijk}\}$) is approximately $N(e_0, s_0^2)$ under H_0 and $N(e_1, s_1^2)$ under H_1, where:

$$e_0 = e^0_{111}, \quad s_0^2 = 1/\sum_{i,j}(n_{ij}p^0_{ij1}p^0_{ij2})^{-1} \tag{6}$$

$$e_1 = e_{111}, \quad s_1^2 = 1/\sum_{i,j}(n_{ij}p_{ij1}p_{ij2})^{-1}. \tag{7}$$

If the normal conditional test rejects H_0 for $|x_{111} - e_0| > s_0 z^{\alpha/2}$, the conditional power will be

$$\tau_3 = 1 + \Phi\left((e_0 - e_1 - s_0 z^{\alpha/2})/s_1\right) - \Phi\left((e_0 - e_1 + s_0 z^{\alpha/2})/s_1\right), \tag{8}$$

and it may serve as an approximation to the unconditional power as well.

4. Comparison of the Three Methods

To compare the performances of the three approximation methods we first generated 40 arrays of "true" probabilities $\{p_{ij1}\}$. Twenty-five arrays correspond to the case $\Psi_1 = 1$, $\Psi_2 = \frac{1}{4}$ $(\vartheta = 4)$ and the other 15–to the case $\Psi_1 = 4$, $\Psi_2 = \frac{1}{4}$ $(\vartheta = 16)$. For each selection of Ψ_1 and Ψ_2, each of p_{111} and p_{211} assumed the values 0.1, 0.3, 0.5, 0.7, 0.9. (For $\Psi_1 = 4$, $\Psi_2 = \frac{1}{4}$, only 15 out of the 25 possible arrays were actually different). The exact power was estimated for each of the 40 arrays by simulating 1000 samples and determining the frequency of samples for which H_0 was rejected at the 0.05 and 0.01 significance levels. This process has been repeated for 11 configurations of sample sizes (n_{ij}).

The observed rejection rates varied considerably for different probability arrays, even with the same Ψ_1 and Ψ_2. Thus, the following criterion has been used to assess the accuracy of an approximation:

$$\delta_A = |\tau_A - \tau_S|/\varepsilon_S, \tag{9}$$

where τ_S and τ_A are the exact (simulated) and approximated powers and $\varepsilon_S = (\tau_S(1 - \tau_S)/100)^{1/2}$ is a standardization factor. An approximation yielding a standardized deviation of $\delta_A \leqq 1$ would usually be considered fair. (Note that ε_S is proportional to the simulation standard error of τ_S.)

Table 1 presents the means of δ_A for the three methods introduced in section 3 as functions of ϑ, α and the sample sizes.

Table 1
Mean of δ_A–equation (9)–for three approximation methods

sample sizes					$\alpha=0.05$			$\alpha=0.01$		
n_{11}	n_{12}	n_{21}	n_{22}	ϑ	LOR	$NC\chi^2$	ACD	LOR	$NC\chi^2$	ACD
10	10	10	10	4	.97	.96	.65	.68	.67	.53
				16	1.11	.78	.69	1.03	.75	1.01
20	20	20	20	4	.74	.71	.47	.77	.78	.41
				16	1.63	.67	.53	1.40	.61	.67
50	50	50	50	4	.32	.24	.28	.55	.49	.27
				16	2.12	.66	.45	2.12	.70	.36
100	100	100	100	4	.40	.28	.23	.39	.37	.29
				16	2.29	.56	.78	2.93	.33	.82
10	10	10	20	4	.52	.49	.35	.56	.54	.33
				16	1.26	.75	.73	.89	.54	.80
10	10	20	20	4	.67	.61	.39	.63	.59	.31
				16	1.24	.47	.41	1.15	.69	.66
10	20	20	20	4	.85	.82	.47	.86	.87	.43
				16	1.43	.42	.35	1.71	.72	.54
10	10	10	100	4	.69	.59	.47	.87	.82	.37
				16	1.38	.59	.61	.81	.50	.48
10	10	100	100	4	.94	.74	.34	.91	.75	.36
				16	2.16	.27	.26	2.07	.35	.39
10	100	100	100	4	.87	.63	.49	1.82	1.67	.43
				16	1.41	2.92	.55	2.17	2.16	.47
10	20	50	100	4	.66	.49	.33	.85	.77	.32
				16	1.92	.65	.28	2.68	.82	.41

5. Conclusions

The results of the present study suggest that accuracy of all three methods is no affected by the significance level of the test. The approximations are generally better for a moderate interaction effect ($\vartheta=4$) than for an extreme effect ($\vartheta=16$).

A comparison of the three methods clearly indicates that the ACD method is preferable to the other two methods, while the LOR approximation is the worst. It is interesting to note that the ACD method, with a slight modification, also provides close approximations to the power of the exact test for the same null hypothesis (HABER, 1983).

References

BARTLETT, M. S., 1935: Contingency table interactions. J. Roy. Statist. Soc. Supp. 2, 248–252.

FIENBERG, S. E., 1980: The Analysis of Cross-Classified Categorical Data. MIT Press, Cambridge, Mass.

GODAMBE, A. V. and HARKNESS, W. L., 1975: Normal approximation to the distribution of a cell entry in a $2\times2\times2$ contingency table. Commun. Statist. 4, 699–709.

GOODMAN, L. A., 1964: Simple methods for analyzing three-factor interaction in contingency tables. J. Amer. Statist. Assoc. **59**, 319–352.
HABER, M., 1981: On the asymptotic power and relative efficiency of the frequency X^2 test. J. Statist. Plan. Inf. **5**, 299–308.
HABER, M., 1983: Sample sizes for the exact test of 'no interaction' in $2 \times 2 \times 2$ tables. Biometrics **39**, 493–498.

Manuscript received: Nov. 11, 1982

Author's address:

Dr. MICHAEL HABER
Department of Biometry
Emory University
Atlanta GA 30322, U.S.A.

Biom. J. 27 (1985) 2, 237–238

On Two Auxiliary Variates in Ratio Method of Estimation

S. K. AGARWAL and PRANESH KUMAR

Mathematics Department, Kenyatta University College, Nairobi, Kenya
I.A.S.R.I., Pusa, New Delhi, India

Abstract

A ratio type estimator using two auxiliary variates is suggested which is found to be more practicable than that of AGARWAL's (1980) estimator.

Key words: Auxiliary variates, ratio estimator.

1. Introduction

AGARWAL (1980) proposed a ratio type estimator ($\bar{y}_w'$) for estimating the population mean $\bar{Y}$ of variate under study y, using two auxiliary variates x_1 and x_2. The conditions were obtained to choose between $\bar{y}_w'$ and OLKIN's (1958) estimator ($\bar{y}_w$) based on two auxiliary variates. The estimator ($\bar{y}_w'$) though more efficient than $\bar{y}_w$ for some populations, may take negative value whenever $\varphi(=\varrho_{01}/\varrho_{02})$, the ratio of correlation coefficients between (y, x_1) and (y, x_2), is either less than ϱ_{12} the correlation coefficient between (x_1, x_2), or $\varphi\varrho_{12}$ is greater than one. The negative estimator of $\bar{Y}$ is meaningless, thus $\bar{y}_w'$ under certain situations has less practical applicability.

The present study deals with another ratio type estimator ($\bar{y}_w''$) which is always positive and gives the same efficiency as that of $\bar{y}_w'$ under the same set of conditions.

2. Bias and Mean Square Error for $\bar{y}_w''$.

The estimator $\bar{y}_w''$ is suggested under simple random sampling with or (without) replacement.

$$\bar{y}_w''=\bar{y}\ (a\bar{X}_1+\bar{X}_2)/(a\bar{x}_1+\bar{x}_2); \quad a=\varphi\sigma_2/\sigma_1$$

where $\bar{X}_i$ and $\bar{x}_i$ are population and sample means for x_i, $i=1, 2$, $\bar{y}$ is the sample mean for study variable y and σ_i; $i=1, 2$ is the standard deviation for ith auxiliary variate. The bias and mean square error in $\bar{y}_w''$, ignoring the terms of order higher than n^{-1}, are given by

(2.1) $$B(\bar{y}''_w) = \bar{Y}\,\{(\varphi^2+1)\,(1 - D\varrho_{02}C_y) + 2\varphi\varrho_{12}\}\,D^{-2}\,n^{-1}$$

and

(2.2) $$\mathrm{MSE}(\bar{y}''_w) = \bar{Y}^2\,\{C_y^2 + \{(\varphi^2+1)\,(1 - 2D\varrho_{02}C_y) + 2\varphi\varrho_{12}\}\,D^{-2}\}\,n^{-1}$$

where

C_y is the coefficient of variation of y,
C_i is the coefficient of variation of x_i, $i = 1, 2$,
$D = (\varphi/C_1 + 1/C_2)$.

Under the conditions

$$\varrho_{01} = \varrho_{02} = \varrho$$

and

$$C_1 = C_2 = C_x$$

(2.1) reduces to

(2.3) $$\bar{Y}\,\{C_x^2\,(1+\varrho_{12})/2 - \varrho C_x C_y\}\,n^{-1}$$

and (2.2) to

(2.4) $$\bar{Y}^2\,\{C_y^2 + C_x^2\,(1+\varrho_{12})/2 - 2\varrho C_y C_x\}\,n^{-1}.$$

(2.4) is the same for all the estimators $\bar{y}_w$, $\bar{y}'_w$ and $\bar{y}''_w$ while
(2.3) is identical to bias of $\bar{y}'_w$ and will be less than that of $\bar{y}_w$ if $\varrho C_y/C_x < 0.75 + 0.25\varrho_{12}$.

References

AGARWAL, S. K., 1980: Two auxiliary variates in ratio method of estimation. Biom. J. **22**, 569–573.
OLKIN, I., 1958: Multi-variate ratio estimations for finite populations. Biometrika **45**, 154–65.

Manuscript received· Sept. 23, 1982

Author's addresses:

Dr. S. K. AGARWAL
Kenyatta University College
Dept. of Mathematics
P. O. Box 43844
Nairobi, Kenya

PRANESH KUMAR
I.A.S.R.I.
Pusa, New Delhi
110012, India

Biom. J. 27 (1985) 2, 239–240

Book Review

FELSENSTEIN, J. (Ed.): Numerical Taconomy. NATO ASI Series. Series G: Ecological Sciences No. I. Springer-Verlag, Berlin – Heidelberg – New York – Tokyo 1983, Proceedings Bad Windsheim, Germany, July 4–16, 1982, 644 pp., DM 149,–.

Numerical taxonomy (NT) was and is a matter in dispute. Is this book a justification? There is no need for that. It is given an analogy. EINSTEIN (according his opinion theories are invented!) wrote in his famous Princeton lectures that in the theory of relativity (RT) the light propagation is excellently important because we know this phenomenon better than other phenomena which are also available to find relations between different space points for "constructing" (defining) the term of time. On the other hand numerical taxonomists are seeking for mathematical or numerical relations between organisms to base classifications on taxonomic characters. Nothing is more sure than the knowledge of characters for constructing classification systems (these are also invented!) because the mankind collect data concerning that subject since millenniums. Thus we find the "same" situation for RT and NT. However in the NT we have no experimentum crucis. In the phylogenetic systematics there are no experiments at all (but comp. the very interesting contribution by R. R. SOKAL on the Caminalcules, "artificially evolved organisms" and "created as animals in 1965" and its cladistic and phenetic classification approaches). Also there exist no "proofs" regarding phenetic classification systems. Therefore it is a very great merit that in approximately 30 (altogether 70 papers in English and I paper in French) the participants of this symposium deal with the basic questions of the (numerical) taxonomy characterized by a high general statement and by a high personal engagement. With conscientiousness and exactness the question concerning a natural or true or general system of the organisms is investigated (to, console taxonomists: each branch, e.g. chemistry, nuclear physics, metal physics and so on, has its own periodic system of the chemical elements). One of the nestors of the NT, P. H. A. SNEATH, speaks on "Philosophy and methods in biological classification" drawing up from a large "pool" of knowledge and of experience. But also "novices" present a lot of many new ideas far from that to be a "re-hash of the old arguments" (J. MCNEILL) or to understand the taxonomy as "classification by an undefined but gentlemanly set of rules" (W. W. MOSS). On the contrary, on the one hand, in spite of large difficulties notions and conceptions concerning performance criteria of a classification are developed and discussed: degree of reflection of phylogenetic relations, degree of reflection of relations of similarity, predictivity of a classification, stability through changes in varying algorithmic steps of similarity procedures, information content, congruence on addition of characters, congruence on addition of OTUs (operational taxonomic units). On the other hand, an important characteristic of these procedings are the many examples presented in all possible dimensions concerning methods (approximately twenty different variants using frequently different mathematics from cluster analysis, theory of graphes, statistics, probability theory, theory of matrices, theory of sets and so on) and objects (organic macromolecules, bacteria, plants in about 10 papers, animals in about 10 papers, human races, ecological items, behavioral variables, organoleptic NT of Portuguese wines). Only in this way the practical verication of the NT can go ahead. In this connection one has to mention the absence of a global theory concerning NT (often apparently general concepts are really imperfect or inconcistent, comp.

e.g. the papers by J. Archie and D. P. Faith refering to cladistic classification). Also for instance, in spite of Shannons definition the information and its optimization in NT is an open question: e.g. 8 species of a genus, u.e. 9 taxa, can yield a classification with a maximum of 15 taxa. What is the optimum between 9 and 15? In connection with such questions the proceedings document both the cowork of biologists and mathematicians and the synthesis of creative and critical elements. At present none knows any phylogeny exactly. However the correctness of the idea of evolution is beyond doubt. None can describe the dangers of a "computer evangelism" (J. Feslenstein). But there is no question concerning availability of electronic data processing. Also the question concerning correctness and availability. of employment of mathematics in taxonomy has such a positive answer as demonstrated by this proceedings. Now the subject ordering organized (even an index) by J. Felsenstein (also 2 large interesting papers) is presented: Approaches to classification, taxonomic congruence, clustering and ordination, reconstructing phylogenies, analyzing morphological variation, geographic variation, biochemical applications general (better: specific) applications, computers and systematics. In conclusion the papers by I. C. Lerman (on some mathematical backgrounds of the automatic classification) and by J.-M. Hubac (on mathematical shape analysis) should be mentioned explicitly. It is impossible to cite the many other interesting papers. In summary one can say: This book could be a jubilee one on the occasion of the twentieth birthday of the NT (in 1963 "The principles of numerical taxonomy" by Sokal and Sneath came out) and represents a further progress in this interdisciplinary science.

D. Leuschner

Molecular Genetic Information Systems Modelling and Simulation

Herausgegeben von KLAUS BELLMANN

1. Auflage
1983. 328 Seiten – 115 Abbildungen – 23 Tabellen – Leinen
DDR 50,– M; Ausland 60,– DM
Bestell-Nr. 763 010 6
Bestellwort: Models Genetic 6647

In 9 Einzelbeiträgen werden deterministische und stochastische, flexible Simulationsmodelle verschiedener Abstraktionsebenen mit Anwendungen auf reale pro- und eucaryotische molekulargenetische Systeme (Qβ, Lambda, Rattenleberzellen) vorgestellt und elementare molekulargenetische Prozesse in Netzen gekoppelter Genexpressionseinheiten betrachtet. (Transkription, Translation, Induktion, Suppression). Diese Modelle beinhalten auch jene strukturellen (baulichen) Charakteristika der genetischen Informationsbestände, die die Dynamik der Transkription bzw. Translation beeinflussen. Zur Lösung der hochdimensionalen, nichtlinearen Parameteradaptionsprobleme wird ein stochastisches Suchverfahren (Evolutionsstrategie) demonstriert und auf ein Modell der eucaryotischen Transkriptionskontrolle angewendet.

Bitte richten Sie Ihre Bestellungen an eine Buchhandlung

AKADEMIE-VERLAG · BERLIN
DDR-1086 Berlin, Leipziger Str. 3–4

BIOMETRICAL JOURNAL

VOLUME 27 · 1985
NUMBER 2

Journal of Mathematical Methods in Biosciences

CONTENTS

Biometrical Journal is indexed in Current Contents/Physical, Chemical & Earth Sciences, and the Science Citation Index.

Bei Fragen zur Produktsicherheit wenden Sie sich bitte an:
If you have any questions regarding product safety,
please contact:

Walter de Gruyter GmbH
Genthiner Straße 13
10785 Berlin
productsafety@degruyterbrill.com